THE ESSENCE OF TIME

Novels by Joseph P. Cody

THE TIGER'S FURY

DRAGON FANG

HUBOT - Human Robot

METANOIA - Total Conversion

PHANTOM TEAM

WILD VIOLETS - Growing Up
In The 1940s And 50s

FIND SPARTACUS

Non-fiction by Joseph P. Cody

THE END OF THOUGHT

THE ESSENCE OF TIME

JOSEPH P. CODY

The views expressed herein are solely those of the author.

This book is written, printed, and bound in the United States of America

250715-1

ISBN: 978-0-9791167-0-4

A Publication of:

Autotech Industries
688 – 11th Avenue NW
St. Paul, Minnesota 55112

This and other books by Joseph P. Cody may be ordered from Amazon.com or any book store.

To Eric

Contents

Contents

Chapter 1

Time According to Aristotle, St. Thomas and St. Augustine

Introduction

Time is a fundamental part of what it is to be human. Yet we seldom ask what time is, what regulates its flow, or what is our relation to it. In the pages that follow we'll find that the spirit of man is connected to time in an unexpected way. Not only are the soul and time linked but there is the human body to consider which in turn is totally united to the soul. Since the human body is part of the material universe that also becomes involved since time works for material things beyond the body as well.

Here it will be necessary to look at many aspects of time so the reader can grasp its real meaning. As such, a few words must be said about the references cited in what follows. Normally someone reading about time would expect the writer to stick to uncontroversial scientific sources. As will be shown that wouldn't get us far. An honest inquiry into a topic as ethereal as time would demand that the writer use all the information available regardless of its source. As a result we will go beyond strictly science and use philosophical sources, and yet further afield and consider Divine Revolution. Some readers may object to the later two sources as being unsuitable especially Divine Revelation.

It is well to reflect on how we have let our thinking become so narrowly constrained. In the sixteenth and seventeenth centuries social and political change along with economic development led to the rise in the demand that the only valid knowledge came from the rational study of nature through observation and experimentation. The rising insistence that science, as the self-proclaimed and exciting engine of progress, must sever all connection with revelation and in no small part from philosophy came into vogue. Prior to this, what we call science was included in the general word philosophy which dealt with the study of all things naturally knowable. And that is precisely why we don't understand time as well as much else in the world around us. An attempt will be made to remove the blinders placed upon our thinking those centuries ago and see a wider world. With that said it is hoped that the following will be taken as a whole so the reader absorbs an appreciation for a subject that is rarely mentioned.

Essence

Since this is to be a study of the essence of time it is well to start with a few words about the meaning of essence. In this regard there are two related concepts to be distinguished—essence and being. On a fundamental level one can say that "the essence is what is signified through the definition of a thing." [1] And a being is an exemplar of that definition or essence. In the real world of composite substances like rocks and people a composite substance is composed of matter and form. In people the soul is what forms the primary matter into us. "Matter in this context is the primary matter that combines with substantial form to constitute a sensible substance. In itself it [primary matter] is not actually something or knowable. It is the pure possibility of receiving form." [2] In the case of substances "essence embraces both matter and form." [3] Or we can say that existence actualizes the possibility within the limits imposed by the distinctive nature of the essence thus made actual.

However, time is not a composite substance of the type stated so we must look further. There is a category of existing things that do not contain matter and these are spirits of which more will be said later. These come in three types, the infinite spirit, namely God, angels, and human souls. Time certainly doesn't fit into any of these three categories except perhaps God since he is "all in all." There is another possibility and that is the idea of accidents as essences.

"Because essence is what the definition signifies, as has been said, accidents must have an essence in the same way that they have a definition. Now their definition is incomplete, because they cannot be defined without including a subject in their definition." [4] Or more succinctly, "An

---

[1] Thomas Aquinas, *Thomas Aquinas On Being and Essence*, translated by Armand Maurer, 2nd Edition, 1968, The Pontifical Institute of Mediaeval Studies, Toronto, Canada, Chapter 2, § 1, p 34.
[2] Ibid.
[3] Ibid., p 35.
[4] Ibid., p 66.

accident is an essence whose nature it is to exist in a substance as in a subject." [5]

That leaves the interesting prospect that the essence of time is an accident of human beings that are alive. If it can be shown that time only exists for humans that are alive that would be a reasonable statement. In what follows an effort will be made to show that statement is true.

Aristotle and St. Thomas Aquinas

Normally one would start this study with Aristotle. Here St. Thomas Aquinas has been included because his thinking on time follows almost exactly that of Aristotle. Without having to get too involved in either Aristotle or St. Thomas we will start by making use of a book by H.D. Gardeil where his section on time has many references to both philosophers. In his section on *The Nature of Time*, Gardeil says:

> But, though time is not motion, it is nevertheless unseverable from motion. Take away all change or motion, and time disappears. That is why the awareness of time dies with the awareness of change, as happens in sound sleep. No motion, no time, so much is true. But, motion is time, no. Hence, though not identical with it, time is yet somehow affiliated with motion. What is the affiliation? What, in other words, is time?
>
> Aristotle's answer is progressive. Time, he says, is continuous; it attends motion, and motion implies extension, which is continuous. This, then, is one thing that defines time, it is continuous. Secondly, in magnitude there is a before and after, namely, of position. . . . But, note, thirdly, what we do when we perceive a before and after. We distinguish phases of the motion, marking off, mentally, one part from another. That is, we perceive the motion as measurable and numerable, and number it. To differentiate within a quantity or magnitude is equivalent to numbering. In general, therefore, we may say that motion plus numbering equals time a thought that St. Thomas sets forth as follows; "Since succession is found in all motion, and one part follows another, in numbering the before and after of motion we apprehend time, which is nothing else than the number of before and after in motion." [6]

---

[5] Ibid.

[6] H.D. Gardeil, O. P. *Introduction to the Philosophy of St. Thomas Aquinas*, Vol. II, Cosmology, B. Herder Book Co. English translation 1958, pp. 121-2

Here we have a definition of time but a definition does not always tell us what a thing really is. In the following section on *The Reality of Time* Gardeil attempts an answer.

> To know the definition of time is one thing, to know what sort of reality it is may be quite another. So evanescent is time that the question may well be asked with Aristotle whether it has any objective existence at all. Can a thing be real if its parts do not really exist? Yet, time appears to be made of parts that to not exist. It has past and future, but the past is no more, and future is not yet. True, there is also the present moment, but that alone, whatever its actuality, does not constitute time. Add to this that time, it seems, can hardly exist without a mind to piece the parts together. Time is the number of motion. But, without something that can count there should be no number. Yet only an intellect can number. It seems, then, that without a soul (in the sense of intellect) there could not be number, hence no time. [7]

From this one would conclude that time is strictly a subjective thing. That is not strictly true as we see further on in Gardeil as he continues to make use of Aristotle and St. Thomas.

> Aristotle allows that time in its full meaning cannot exist apart from mind. . . . Nevertheless, time is not a sheer subjectivity. The mental work of discriminating before and after and relating them to each other has an objective foundation, being grounded in the motion of which before and after are parts. Granted that motion is an imperfect reality. . .it is nevertheless a reality. Thus, speaking of the objective and subjective elements of time, St. Thomas writes: "That part of time which is as it were its *material* element, namely, the before and after, is founded in motion; but the *formal* element is completed in the soul's activity of numbering. And that is why the Philosopher says that without a soul there would be no time." Emphasis added. [8]

Here St. Thomas makes time a substance by saying it is composed of

[7] Ibid. p. 123
[8] Ibid. pp. 123-4.

matter and form. We will have to look at that more closely because the material of time being made of the before and after is made of nothing because the past is no more and the future is not yet. And, he leaves out the present entirely.

Another modern author is Paul Glenn who has the following to say about time in which he also relies on both Aristotle and St. Thomas Aquinas. "Time is an accident (like position, or action) which determines a reality in its position with reference to before and after. Examples: at midday, this evening, at five o'clock, next Tuesday, in 1492, before midnight, after supper." [9]

He goes more in-depth later as he says:

Bodies with quantity are subject to *change*. Change is movement or motion, for "change is a transit, a going-over, a movement from one state of being to another." Now, movement or motion is a matter of "now this—then that"; it is a matter of "before and after." And motion or change, under the aspect of before-and-after, is the basis of *real time*. Time in itself is <u>described</u> as a continuous and numerable series of motions under the aspect of before-and-after. Man <u>conceives</u> of time as a *measure*, just as he conceives of space as a *container*. But just as space in its reality is the real extension of bodies, so time in its reality is the continuous numerable succession of bodily movements (of sun, of stars, of moon). Time *as a measure* <u>is</u> logical being, [10] not real being. . . . Besides real time we have *ideal time* which is the mind's concept of all possible numerable and continuous movement; and we have *imaginary time* which is the fanciful envisioning of real time indefinitely extended. Real time is necessarily finite, for it is finite motion in a finite world of finite bodies. Ideal time and imaginary time are *indefinite* or *potentially infinite*, but never actually infinite. Thoughtless people sometimes confuse ideal or imaginary time with *eternity*. But eternity is, strictly speaking, the opposite of time. It is an endless

---

[9] Paul J. Glenn, *An Introduction To Philosophy*, St. Louis, Mo., B. Herder Book Company, 1943, p. 249.
[10] Logical being is being of reason: that which can only exist in the intellect conceiving it. H. D. Gardeil, O. P., *Introduction to the Philosophy of St. Thomas Aquinas*, Vol. IV, Metaphysics, B. Herder Book Co., 1967, pp. 299.

"now"; it has nothing of "before and after" which is the essence of time. (All emphasis his except underlines which are added). [11]

Notice that in speaking of eternity above Glenn defines eternity as: "It is an endless 'now'; it has nothing of 'before and after' *which is the essence of time*." Emphasis added. In so doing he defines the essence of time as "before and after." Above he also says time is an accident so he seems to be saying that time is an accidental essence consisting of "before and after." We will revisit this later.

There are some additional things one finds interesting about Glenn's treatment of time. The first is that it is all in terms of "before and after," since he never mentions the present. Aristotle, St. Augustine and St. Thomas dwelt on before and after, too, but also considered the present. The closest Glenn comes to mentioning the present is in the definition of eternity which is and endless "now."

The second point is that he uses "numerable" to describe time. This again is in accordance with the others mentioned. From this we could say that time is the number of motion, according to a before and after. But, that describes what we *do* with time not what time *is*. It is like asking what air is and someone saying that air is the gas we breath in and out to stay alive. That statement tells want we *do* with air not what air *is*.

Gardeil refers briefly to the present as follows: "Can a thing be real if its parts do not really exist? Yet time appears to be made of parts that do not exist. It has past and future; but the past is no more and the future is not yet. True, there is also the present moment, but this alone, whatever its actuality, does not constitute time." [12]

One could argue that the present is the only time that does exist because as they all agree, the past and future do not exist. Since we do exist we exist only in the present so that part of time *must* exist and represents the only real part of time. As Glenn says above, "Time is an accident (like position, or action)." Whether or not time is an accident is the question before us. But, more importantly he associates time with "position" as in space or "action" which is motion which in turn is the change of position in space, and time is used to determine motion. That means he is using time to define time. In any case, all of this measuring of before and after only describes what time is used for not what it is.

---

[11] Ibid., Glenn, pp. 263-4.
[12] Ibid., Gardeil, Vol. II Cosmology, p. 123

Perhaps what the philosophers do not make clear is they are looking at time as an accident. In addition to position and action, color, for example, is an accident. All of these accidents are part of physical beings. Attaching time to a "thing" as an accident is like attaching the thing to space. Of course, a "thing" needs space to exist just as it needs time. But, we do not think of space as being an accident because all things need space and accidents are what distinguish one "thing" from another. In like manner, all things need time.

An accident does not exist by itself but only in a thing. There is no such thing as yellow. It must be a yellow flower or a yellow button. However, yellow does exist in our minds apart from any specific thing. And, one could say that time does exist in our minds apart from any specific event. The difference is that the only place that time really does exist is in our minds since the future is not yet, the present has no length and the past is no more.

St. Augustine

To take this a step further, we are indebted to St. Augustine for one of the best discussions of time in his *The Confessions of St. Augustine*. He lived from A.D. 345 to 430 and wrote this work in the years around 400 so he was not yet fifty. In Book 11 of that work, he is pleading with God to give him insight to this vexing problem of time. He considers the question of the beginning of the world and with it the beginning of time and shows that time and creation were made at the same time. But what is time? To this Augustine devotes a insightful analysis of the subjectivity of time and the relation of all temporal process to the abiding eternity of God. Chapters I through XIII of Book 11 are not particularly relevant to our purposes here, though the reader is encouraged to read them. As a result we start with Chapter XIV. Here we are using a translation by Albert C. Outler, Ph.D., D.D. It uses contemporary language and it is well referenced. As one might expect there are many translations of such a famous work which was originally written in Latin. As such there are times where the interpretations vary.

To summarize, then, though inadequately, St. Augustine looks at time future, present, and past arriving at the conclusion that the future does not exist because it is not yet, the present does not exist because it has no length, and the past does not exist because it is no longer. Finally he arrives at the point that the only place time, especially future and past, exist is in the human mind. His conclusions are the same as those of Aristotle and St. Thomas. It is not clear if he had access to Aristotle's work on time. Of course, St. Thomas was still centuries in the future when St. Augustine lived. At times his treatment is perceptive, at other places he stops and appeals to God to give him the light to understand his topic.

To give the reader a taste of the work Chapter XIV will be reproduced here followed by parts of other chapters.

Chapter XIV
There was no time, therefore, when thou hadst not made anything, because thou hadst made time itself. And there are no times that are coeternal with thee, because thou dost abide forever; but if times should abide, they would not be times.

For what is time? Who can easily and briefly explain it? Who can even comprehend it in thought or put the answer into words? Yet is it not true that in conversation we refer to nothing more familiarly or knowingly than time? And surely we understand it when we speak of it; we understand it also when we hear another speak of it.

What, then, is time? If no one asks me, I know what it is. If I wish to explain it to him who asks me, I do not know. Yet I say with confidence that I know that if nothing passed away, there would be no past time; and if nothing were still coming, there would be no future time; and if there were nothing now, there would be no present time. But, then, how is it that there are the two times, past and future, when even the past is now no longer and the future is now not yet? But if the present were always present, and did not pass into past time, it obviously would not be time but eternity. If, then, time present—if it be time—comes into existence only because it passes into time past, how can we say that even this is, since the cause of its being is that it will cease to be? Thus, can we not truly say that time is only as it tends toward nonbeing? [4]

Chapter XV
Thus it comes out that time present, which we found was the only time that could be called "long," has been cut down to the space of scarcely a single day. . . .And that one hour itself passes away in

---

[4] Albert C. Outler, Ph.D., D.D., *The Confessions of St. Augustine*, Philadelphia, Westminster Press, 1955.

fleeting fractions. The part of it that has fled is past; what remains is still future. If any fraction of time be conceived that cannot now be divided even into the most minute momentary point, this alone is what we may call time present. But this flies so rapidly from future to past that it cannot be extended by any delay. For if it is extended, it is then divided into past and future. *But the present has no extension whatever*. Emphasis added.

Chapter XVIII
Give me leave, O Lord, to seek still further. O my Hope, let not my purpose be confounded. For if there are times past and future, I wish to know where they are. But if I have not yet succeeded in this, I still know that wherever they are, they are not there as future or past, but as present. For if they are there as future, they are there as "not yet"; if they are there as past, they are there as "no longer." Wherever they are and whatever they are they exist therefore only as present.

Chapter XX
But even now it is manifest and clear that there are neither times future nor times past. Thus it is not properly said that there are three times, past, present, and future. Perhaps it might be said rightly that there are three times: a time present of things past; a time present of things present; and a time present of things future. For these three do coexist somehow in the soul, for otherwise I could not see them.

Chapter XXVIII
But how is the future diminished or consumed when it does not yet exist? Or how does the past, which exists no longer, increase, unless it is that in the mind in which all this happens there are three functions? For the mind expects, it attends, and it remembers; so that what it expects passes into what it remembers by way of what it attends to. . . . [5]

---

[5] Ibid.

Chapter 2

Other Thoughts About Time

As one would expect there are about as many ideas as to what time is as there are people. On the Internet one can find some thoughtful as well as downright crazy ideas in this regard. That being so this chapter will hit a few highlights and make no attempt at being exhaustive.

The Human Meaning of time

To be human is to have a deep sense that something is missing, that we are lacking something. This occurs on a spiritual level that can be seen in two forms. One might call the first a direct longing which was probably said best by St. Augustine, "Our souls are restless, Lord, until they rest in Thee." The second is a response to an emptiness that we can, at least partially, satisfy by having a connection with the past and the future. This is commonly done by forming families, though that is not the only way. Certainly it is not proposed that it is necessary for every man and woman to marry and have children, just that societal structures must be oriented in a direction that those who do have families have the support of the wider community to be successful.

Through fertility time becomes a continuity as each person sees himself as having come from the previous generation, his parents, and children as connecting him to the future. This is why we find a need to connect with our past, why we need a history. Beings that do not have this basic, supernatural need do not have a history. Animals are fulfilled in that they know what to do as dictated by instinct. Humans do not know what to do and must be taught. And what must we be taught? It is how to do the things our ancestors learned that will help us survive; it is the passing along of culture, of making monuments, of learning our history. This man of time, of history, embodies in himself his tribal past. He has an impetus to create and sustain institutions that best permits heirs. In the modern, liberal culture our history is being rewritten, lied about, torn down and denied; our monuments are being destroyed.

We see it in the destruction of the Civil War memorials in the South. We see this in the way schools denigrate the founding fathers of our country. We have even seen it as the liberals have taken it upon themselves to rename one of the lakes in Minneapolis—Lake Calhoun. They have replace it with Bda Maka Sha meaning Lake White Earth supposedly an "historical" Indian name. Interestingly, the Indians also called it Mde Medoza, Loon Lake, which was the name adapted by settlers, or Heyate Mde, Set Back Lake. But, Calhoun *is* an historical name that connects to *our* history. Further, it is unlikely that any Indians in the metro area knew or cared about the new name until a white liberal brought it to their attention. That does not matter because the object is to erase *our* history. The overriding goal of all of this is for every man to become an island, the self-sufficient being whose time has collapsed into nothing leaving him on the same level as an animal—a beast.

This state of being can be seen as a contrast between space and time. Animals are not aware of time; they live only in space. Self-sufficient man orients himself to space, not time. Having no connection with the time of humanity he cares little about history or successors but only about controlling and inclosing his life in a personal space or extending it to control others through war or debt. Having no successors this space oriented man has no future. This is seen most profoundly in the dearth of births in Western Civilization as the time oriented man of Christendom has become a space oriented man of modernism.

Time as Seen by Scientists

Science as it is practiced today assumes a totally determinate universe that started with the big bang and will continue along for more billions of years. A scientist in his lab sets up an apparatus to study a small part of the physical world. He makes a measurement and records it. By the time the datum is entered in his ledger the part of the physical world from which he took the measurement no longer exists. It is in the past. But, he is banking on the fact that the universe is determinate and that the part of the physical universe now present in his apparatus resembles the one he measured in the recent past closely enough to make his observation predictive. That is to say, he is only concerned about time as a numbering of motion. His clocks work and that makes sense of his measurements. There is nothing wrong with that; that is how we lead our lives.

Time as Seen by the Cyclic Creeds

The cyclic creeds such as Buddhism and Hinduism do not include telos in their theology and as a result there is no direction or goal to existence. That is, everything keeps repeating itself. "In the scheme we have set out above [concerning the cyclic creeds] the unfolding of time is a development without substance in which nothing changes because everything changes." [1] This is another case where the refusal to accept authentic divine revelation hampers ones ability to adequately comprehend reality. For reference, Buddhism began in the sixth century BC and Hinduism was a later development of elements of Buddhism and other sources.

World Lines

George Gamow proposed something he called world lines such that each physical thing in the universe has one. By this he means that a thing has a location not only in the common three dimensions of space, but also a location in the forth dimension of time. In that case everyone has a unique world line that goes weaving through this strange space-time continuum. "We see that from the point of view of the four-dimensional space-time geometry the topography and the history of the universe fuse into one harmonious picture, and all we have to consider is a tangled bunch of world lines representing the motion of individual atoms, animals, or stars." [2] For man his world line starts at his conception, and ends when he dies. The interesting thing about this for believers is that everyone's world line ends at exactly the same place, namely their personal judgement before the throne of God. Everyone is given an allotment of time in how long they live and we each must determine what we do with it.

Time Travel

Can there be time travel logically irrespective of the physical possibility? The answer would have to be "no" to travel to the future based of the

---

[1] Henni de Lubar, *Catholicism, Christ and the Common Destiny of Man*, London, Bruns and Oats, 1950, p. 141.

[2] George Gamow, *One, Two, Three. . . Infinity*, New York, Mentor Books, 1947, p. 78.

scriptural references that say the future is known only to the Father. This is also reasonable. The future is determined by the free will of man acting in the present. With all the possible free will decisions that all the people alive at any one time make from instant to instant, the future is far too complicated for anyone but God to know.

As for actual physical time travel, let us start with travel to the past. If one were to do that it would be an event in the lives of the people before now that would change what they did, namely the decisions they made. That would have a ripple effect through all the time from then to the present and into the future. So the present we would have would not be the present we actually do have.

Next, try travel into the future. If it did happen it would mean that the time traveler could not return to the present because if he did it would amount to the same thing as traveling to the past. Since the future is the result of everything that has happened up until the present and all the free will decisions of man and the calamities of nature that will happen to the point in the future where the time traveler landed, he would be one more calamity for the people of that time to deal with. And since the future has not happened yet, and will be determined by countless second to second free will decisions of billions of people, there would be nearly an infinite number of futures into which our traveler could go so the one our time traveler went into would be most unlikely to be the one that actually happens.

Would it be possible to time travel only to look at future times, and not actually be there to partake in the events of the time? The answer to that should be "no" because knowledge of the future, would change decisions made now and the future would be different than it would be without that knowledge. But, does that matter since the future is indeterminate anyway?

Looking at the past we would have to assume that as each *present* disappeared into the past it would stay intact like a frame in a motion picture. That could be, but there is no reason to suppose that is what happens. If that is not the case the past is completely lost to us.

How Long Is the Present?

As we have seen above St. Augustine arrives at the conclusion that the present has no length. This is one of those questions that people rarely think about. Regardless of the length of the present, time appears to flow at a rate that we can for the most part keep up with. After all, *it is our reality*, so why would God make a reality in which we could not function. Think about a speeding bullet. We cannot see it. But a jet fighter plane can fly as fast as a bullet and we can see it. That is first of all because the fighter plane is much larger and further away. And second, there is nothing commonly in nature that is as small and as fast as a bullet that is dangerous to us or that we care about for any other reason. As a result, time moves and our senses perceive at a rate that is compatible with normal life on the average letting aside that time flies when you are having fun and drags when you are in pain.

Getting back to the speeding bullet, we do not see it because the rate of travel of its image across our retina is faster than the image of the jet plane. Apparently there is not time for neurons to fire and send a signal to the brain that something is there in the case of the bullet. Or else, the signals are in fact there, but we do not consciously recognize them. Our concept of time, the only real time, is regulated at a speed that takes into account the physical processes such as the speed of light being very fast, the size of a photon relative to a rod or cone on the retina, and our brain's rate of processing phantasms. Since all non-spiritual concepts come first by way of sense images that cause a phantasm in our physical brains, and since these images are formed at a rate consistent with physical processes, the two are intertwined. That is, there is a harmony between the speed at which physical processes operate and our ability to process sense images. Here we see that the flow of time is keyed to living things in a way that makes sense. If we assume God wanted to make us, and He obviously did, the demands of nature, as mentioned below, would say he had to make the rest of creation to compliment us.

All of this still does nothing to answer the problem of putting a length of time on the present. Clearly the present has something to do with consciousness, and only humans have consciousness. One must put aside sentient consciousness, that of animals, which is another thing entirely. Modern writers like to confuse the two so they do not have to deal with a spiritual soul in man. We know about the present but only through a glass darkly. Time is mostly about the future and the past. Think about making a fist with your hand. First you think about doing it and realize you do know how to do it. You do it and then you have a fist. It happened but then the act is gone, it is in the past. It is gone forever. In fact, if you move your fingers really fast, you hardly see them move in the act of making a fist, especially if your fingers are close to your eyes—the speeding bullet effect.

But, we are still no closer to the length of the present. One thing is certain, it is not long. All of our presents are out there in the future coming toward us with the speed of time. The odd thing is we have some control over those presents coming toward us from the future by how we conduct the present present. If we decide to go shopping this afternoon it will make a different set of presents coming at us from the future than if we stay home. If we knew how many presents there were in a second we would be able to tell how long one was. Maybe they are of all different lengths or maybe they have no length at all as St. Augustine surmised.

Length of the Present as Seen From Theoretical Physics

There is a "constant" that appears in theoretical physics and quantum mechanics that might suggest how short the present could be if it were not zero. That is the Planck time = $(hG/c^5)^{½}$ = 1.35 x 10^{-43} second. This time is based on Planck's constant, h = 6.63 x 10^{-34} joule-seconds, the gravitational constant, 6.67 x 10^{-11} m^3 kg^{-1} sec^{-2} and the speed of light, 3 x 10^8 meters/sec. [3] Planck's constant has been considered a building block for quantum mechanics and theoretical physics for a century and the gravitational constant as well as the speed of light have been measured to considerable accuracy. A length of time of 10^{-43} seconds is unimaginatively tiny. Some have suggested that the shortest time physically possible would be set by Planck time and therefore further suggest that the length of the present would not be zero. However, if that were the case we run into the problem discussed by the ancients namely that if it is not zero, it can be divided into part past and part future. As we shall see later, having the present of no extension at all is the better option.

The Starting and Ending of Time

According to modern science the universe was created (they would say *began*) 13.7 billion years ago with the big bang. But, if it were that long ago or one microsecond before the first rational human mind was created it would not matter. That is because before the first rational mind was created there was no *reason* for time to exist. Animals do not have spiritual souls, therefore are not rational and are not aware of time because

---

[3] This information is readily available, cf.: https://www.space.com/what-is-the-planck-time

they are not aware of themselves or anything. Animals are simply not aware. They could just as well have been created in that instant before the soul of Adam was created and it would not have mattered a bit. There is a saying that takes some time to digest when one is presented with it for the first time. It is this: *Time is a physical quantity that is only used cognitively.*

Of course modern scientists could not allow everything to be created in an instant because creation presumes God exists, an idea they exclude from their science, and further it is not as tidy as assuming a long gestation period for things to get to their present state.

Time Ends with Man

Even if we are not certain that time started with the first rational man, Adam, we do know that it will end with the last one. There are many references to the end things in Sacred Scripture, that is Divine Revelation and they are explicit on the fact that there will be people alive at the end of the present universe. As stated before, this author makes no apology for using Divine Revelation as a source of information since to reject it is short sighted. It is only reasonable to use any and all knowledge available. It is hoped the reader will not be unduly put off for this. Plus, in what follows, it will be shown that one can come to these conclusions with natural reason.

We are told that in the end after a great calamity the present world will pass away and the last judgement occurs. We will see the end of the world and the appearance of the new heaven and new earth and the start of eternity. A few citations follow.

29 And immediately after the tribulation of those days, the sun shall be darkened and the moon shall not give her light and the stars shall fall from heaven and the powers of heaven shall be moved. 30 And then shall appear the sign of the Son of man in heaven. And then shall all tribes of the earth mourn: and they shall see the Son of man coming in the clouds of heaven with much power and majesty. 31 And he shall send his angels with a trumpet and a great voice: and they shall gather together his elect from the four winds, from the farthest parts of the heavens to the utmost bounds of them. 32 And from the fig tree learn a parable: When the branch thereof is now tender and the leaves come forth, you know that summer is nigh. 33 So you also, when you shall see all these

things, know ye that it is nigh, even at the doors. 34 Amen I say to you that this generation shall not pass till all these things be done. 35 Heaven and earth shall pass: but my words shall not pass. 36 But of that day and hour no one knoweth: no, not the angels of heaven, but the Father alone. Mat. 24: 29-36. [4]

Looking for and hastening unto the coming of the day of the Lord, by which the heavens being on fire shall be dissolved, and the elements shall melt with the burning heat. But, we look for new heavens and a new earth, according to his promises, wherein dwells justice. 2 Peter 3:12-13.

I saw a new heaven and a new earth. For the first heaven and the first earth was gone: and the sea is now no more. Rev. 21:1.

And God shall wipe away all tears from their eyes: and death shall be no more. Nor mourning, nor crying, nor sorrow shall be any more, for the former things are passed away. And he that sat on the throne, said: Behold, I make all things new. Rev. 21:4-5.

It seems that God the Father has a "time" picked to the day and the hour when the world will end and time will end, and there will still be human beings living when it happens. Therefore, it is wrong to say that at some time in the future due to nuclear war, pestilence, plague, or famine the last human being will die to be survived by ants, cockroaches and rats, and that even after the sun goes supernovae and destroys the earth along with the ants, cockroaches and rats, that the universe will continue on as if on autopilot until it finally runs out of energy and goes away.

Did Time Start with Man?

Thus, having one bracket on time established it is up to us to see if we can establish the other. The easiest way to do that, it seems, is to note that it is only logical that if there is no need for the world (the universe) or time after there are no more people, that there would not have been any need for either before there were any people. This speaks to the theses that before the first man, a spirit in substantial union with what we call physical matter, there was no need for either the physical universe or time.

---

[4] Mk. 13: 24-32 has almost identical language.

Other Thoughts About Time

We are all familiar with the two creation narratives in Genesis, the first chapter of the Bible. In the first there are six days of creation. This account starts with the heavens and the earth existing and the earth being a formless void. Then God created light, day and night, the sky, dry land and the sea, plants, the heavenly bodies, fish, birds, animals, and only then man. The second account differs in interesting ways. In this one God first created the earth and then the heavens. Then he created Adam followed by vegitation, animals and finally Eve. That is, between the time he created Adam and the time he created Eve he created everything except the raw material, so to speak, the earth and the heavens. That is, God does not allow for time before the first human being in this account. Take your pick of the two but they certainly do not disallow that there is the possibility of no time before humans arrived.

Chapter 3

Human Nature

We must ask the question that if we propose that time is specifically for man or even if man could control the flow of time does he have the power to do so? In this we are, of course, speaking of man's soul. It's important to remind ourselves that a spirit is outside of space and time in its existence. That means that normally it is like oil and water, spirit and matter do not mix, except for us in some mysterious way known only to God. Matter and spirit are simply two different realms of existence. So if our spirit could interact with matter beyond animating our body, there's no reason why it could not affect, control or even destroy matter in other ways, or as far as that goes, control time.

Spirit

We will be speaking about spirit so it is well to spend a few words on discussing what, in fact, a spirit is. To start, a soul is the life principle in a living body. Plants, animals and man have souls. Only in man is the soul also a spirit. So what, then, do we mean by spirit? A spirit knows, wills and has power. Our rational, thinking part does not reside in our brains; it is in our spirits. When we say mercy is kinder than justice we are not comparing electrical impulses in our brains. It happens in our souls.

Our ideas are not material. They have no resemblance to our body. Their resemblance is to our spirit. They have no shape, no size, no color, no weight, no space. Neither has spirit whose offspring they are. But no one can call it nothing; for it produces thought, and thought is the most powerful thing in the world—unless love is,

21

Joseph P. Cody

which spirit also produces. [1]

A somewhat difficult idea about spirit compared to material things is that a spirit has no parts.

> A part is any element in a being which is not the whole of it, as my chest is a part of my body, or an electron a part of an atom. A spirit has no parts. There is no element in it which is not the whole of it. There is no division of parts as there is in mater. Our body has parts, each with its own specialized function: it uses its lungs to breathe with, its eyes to see with, its legs to walk with. Our soul has no parts, for it is a spirit. There in no element of our soul which is not the whole soul. It does a remarkable variety of things—knowing, loving, animating a body—but each one of them is done by the whole soul; it has no parts among which to divide them up. [2]

When we say the spirit has no parts it necessarily means it is indivisible. The reader may think this is a case of belaboring the obvious but it is an important distinction as will become obvious as we proceed.

Since spirit has no parts it does not occupy space since space is what matter spreads its parts into. Having no parts also means a spirit cannot be changed into anything else, nor by any natural process can it be destroyed.

> A spiritual being, therefore, cannot lose its identity. It can experience changes in its relation to other beings—e.g. it can gain new knowledge or lose knowledge that it has; it can transfer its love from this object to that; it can develop its power over matter; its own body can cease to respond to is animating power and death follows for the body—but with all these changes it remains itself, conscious of itself, permanent. [3]

That spirit has power beyond what we find in the material universe becomes apparent only when we look closely at the result of a being that is partless. Having no parts means first that it has nothing in common with

---

[1] Frank J. Sheed, *Theology for Beginners*, New York, Sheed & Ward, 1957, p. 8.
[2] Ibid., pp. 8-9.
[3] Ibid., p. 10.

the material world. It also cannot lose any part of itself the way friable matter can. But more importantly, it is ontologically superior to and more powerful than matter—all matter—due to the total integrity of its being. It takes a fair amount of contemplation for that to finally make sense.

> In our effort to make the concept of spirit wholly our own we must be very leisurely. We must keep on looking at the relation between having parts and occupying space, until we find ourselves quite effortlessly seeing that a partless being is outside space. We must keep on looking at the meaning of being and the meaning of part until, again quite effortlessly, we see that such a being must by its concentration be superior in essence and in operation—that is, in what it is and what it can do—to those others whose being and operation are dispersed. [4]

This might be a good place to interject some thoughts about the difference between imagination and intellect since we live in a spread out material world—the domain of imagination—and our thoughts are entirely in the intellect—the domain of the spirit.

> By positive effort we shall come to realize that our "need" for space is a trick that imagination has worked upon us. Imagination needs space, because it makes pictures, and pictures need space. But intellect does not. Indeed once intellect gets into its own stride, it finds that space does not help but baffles. The philosopher, asking questions, finds space too empty of meaning to be coped with satisfactorily; quite literally it is harder to understand than spacelessness. The reason why imagination can take it so easily is that it asks no questions, but only makes pictures. [5]

The Power of the Human Spirit

In the book *Occult Phenomena*, the author, Alosi Wiesinger, writes about the powers of the spirit we posses as our souls. The reader should understand that in *Occult Phenomena* he looks at "occult" from its real meaning which is "hidden," as when the moon occults the sun shining on the

---

[4] Frank J. Sheed, *Knowing God: God and the Human Condition*, New York, Sheed and Ward, Inc., 1966, p. 74.
[5] Ibid. p. 76.

earth, that is, it hides it, and we have an eclipse of the sun. As such, the book deals with powers of the human soul that are normally hidden and that are only occasionally encountered. This is opposed to the meaning where occult is associated with bizarre cults frequently involving Satanism.

> The faculty of sight, for instance, is in the eyes, but the soul's capacity for cognition is not confined to any one part of the body; indeed in this respect the soul is not only not wholly present in every part of the body, but not wholly present in the body as a whole, for the power of the soul exceeds in its activity the capacity of the body. When therefore I speak of the partly body-free soul, I am not suggesting that there is a substantial separation from the body, but that its purely spiritual powers reach beyond the body's domain and that in this way it is empowered to perform feats in which the body has no part, or simply an abnormal one. [6]

As we see animating our physical bodies is a simple task for the power of a spirit, any spirit including our souls. One of the powers of the soul that each of us has, but rarely think about, is our subconscious. He suggests that our subconscious holds far more memory than our conscious memory.

> Everything that flows towards the soul from the outside first enters into the subconscious, and from here only a small part goes into the upper consciousness at all. The subconscious is therefore much the richer of the two; it leads an independent life, being, so to speak, "busy behind the scenes". It can thus provide an explanation for much that seems to us incomprehensible and surprising. Though everything does not penetrate into the upper consciousness, yet nothing is lost. Experiences may only enter the consciousness after delay, or even not enter it at all, yet they remain effective and condition the freedom of our actions—or they have the effect on us of an alien intelligence. [7]

He also suggests the other occult phenomena such as psychokinesis—the moving of physical objects with the mind—is simply a manifestation of the power of our sprits. His premise is that when these paranormal pow-

---

[6] Alosi Wiesinger, O.C.S.O., *Occult Phenomena*, Westminster, Maryland, The Newman Press, 1957, p. 56.
[7] Ibid., p. 61.

ers are exhibited, they are the peeking out of some to the preternatural gifts that were lost to humans by Original Sin.

According to Wiesinger, prior to the fall, Adam had the powers of a pure spirit.

> By his angelic intelligence Adam knew how to avoid the causes of death and disease and by his will he was able to direct the fluid and solid substances of this world, so that they not only did him no hurt but greatly contributed to his happiness. [8]

He expands on this by saying that our first parents would have had complete power over material substances.

> Our first parents possessed the preternatural gift of a spiritual will which reached out beyond the body, a will which gave man the power of acting on matter and moving it without any kind of effort, even as pure spirits can act upon it and move it. We may thus suppose that Adam performed bodily work for so long as this gave him pleasure and redounded to his health. When, however, it threatened to become wearisome, he used his angelic powers over matter, as he required them. [9]

It seems reasonable that if Adam could control matter, that is move, mold and in some ways change it the same as angels can, he would have had the power to control time as well. Now, angels are not subject to time. "Angels . . . [have] endless duration without the vicissitudes of time." [10] So angels would not have a reason to control time. But, Adam did live in time and if he could control everything else, why not time?

So, in answer to our question, it seems reasonable that before the fall, Adam and Eve would have been able to control the flow of time if there had been any reason to control it. Adam could control matter to avoid diseases and to help with his work when he became weary. There seems no reason for him to control time, though. And even if Adam could control time before the fall, can we collectively as the human race, do it now after the fall?

---

[8] Ibid., p. 84.
[9] Ibid., p. 85.
[10] Ibid., Glenn, p. 264.

Joseph P. Cody

Triune Matter

If all of this seems a little hard to believe, we might as well make it worse by considering one of those unique things about matter that is necessary for us to exist. It is well proven in experimental physics that subatomic particles can have both wave and particle properties. That is, in some experiments they behave like they were particles—matter—and in others like waves—energy. In a similar fashion electromagnetic waves, such as visible light, also appear as waves in some cases and particles in others. This defines a duality that could be said to mean that in some cases these entities appear as matter and in others as energy. Further, as we have seen in atomic power plants and nuclear weapons matter can be converted to energy and vice versa and there is an equivalence expressed by Einstein's famous equation $E = mc^2$. Beyond those two manifestations of matter there is a third.

> The quality of spiritual impulses, ideas, plans, will-power, attitudes, etc., cannot be measured with physical-chemical methods, (even though these ideas, etc. use physical-chemical carriers of information), because these spiritual qualities are more than merely physical-chemical. Reducing reality to physical-chemical appearances therefore, is an arbitrary act of faith which is unusually unscientific. . . .
>
> In spite of the fact that thoughts and impulses of the will cannot be measured it is apparent that the elementary particles—atoms, molecules, cells, and organs of our bodies—respond to the will and spirit, and carry out their wishes. It is also apparent that a [magnetic] tape is not changed quantitatively when it receives a recording, but is electromagnetically changed qualitatively so that it can transmit a new quality. The qualitative power of our elementary particles is possible only when they possess a third form of appearance. A qualitative form, appropriate to the spiritual form, which makes responses to the spirit possible, must exist over and above the electromagnetic quality of a wave or a particle. This is a quality to which the waves or particles can respond. The physicist, Bernard Philberth, explained this type of understanding of living matter, with its three forms of appearances as a "triune," "trinitarian" dimension, in his book *Des Dreieine*. (Bernard Philberth, *Des Dreieine* Christiana Verlag, Stein a. Rhein, 1974.)

> We experience the power of our physical-chemical energies to respond to spirit and will in the activity of our brains. If these energies are capable of reacting to spirit in the brain, they are, in principle, also capable of doing so in other parts of our organism. [11]

The existence of triune matter is seen in practical terms when we, for example, have a thought in our minds that we want to move our hand and then do it. There has to be a connection made from our spirit to the matter or energy in our brains to produce the chain of physical-chemical events that cause the muscles in our arm to contract as directed. This is psychokinesis—controlling or moving matter with our minds—on a micro level. And, we all do it all the time.

This means that we all have the power to move or affect at least a little matter with our minds. As we discussed earlier, the animation of or physical bodies is a ridiculously simple thing for the power of a spirit. Therefore it is not unreasonable that the collective human race plays a substantial role in keeping the physical universe in existence.

God Looks at the Human Race as One Thing

If we do control the flow of time now, would it be done by us collectively, that is all of us alive today, and all who have lived from Adam to those alive at the end of time? One way to answer this is to note that in some mysterious ways God looks at the human race as one thing. This is especially seen in the mysterious idea of original sin. Adam sinned and we all inherit that sin, or one might say that condition, from him.

> Obviously there is something in the solidarity of the whole human race clear to God but not to us, that He could so treat the race as one thing. . . . God sees the whole race, every member of which He created, as one thing—somewhat as we see a family as one thing or even a man. The mere number and variety—myriads upon myriads of men—and the uncountable ages, do not impede the vision of the Eternal and Omniscient God. [12]

There is another way that God looks at the human race as one thing. That is in the general judgement at the end of the world. Why have a General

[11] Siegfried Ernst, MD, *MAN The Greatest of Miracles*, Collegeville, MN, The Liturgical Press, 1976, pp. 148-9.
[12] Ibid., Sheed, p. 57.

Judgement if each one of us is judged as to going to heaven or hell at his personal judgement when he dies? It is to show the absolute justice of God's judgements. We will see everything everyone who ever lived did and how it affected everyone else, especially us. That is, we all affect everyone else. "The Last Judgement will reveal even to its furthest consequences the good each person has done or failed to do in his earthly life." [13]

Concerning the human race as one thing from the perspective of original sin a quote from Wiesinger's *Occult Phenomena* will follow.

Original sin, the sin of our first parents, inherited by all their posterity, consists formally in the deprivation of sanctifying grace with which man had been endowed by God and which he lost both for himself and for the whole human race—as indeed is plainly stated in St. Paul (Rom. 5. 12): "As by one man sin entered into this world and by sin death . . . so death passed upon all men in whom all have sinned." . . .

The difficulty, as it seems to me, is not that all men should be punished. . . in the case of original sin, however, we are not only all punished, but we are all guilty. . . .

The difficulty becomes even greater when the theologians tell us—and quite rightly—that original sin must be for us a free act of the will. . . . It must be a free act of the will if it is to be a real sin at all, even if it is only an habitual state of fallen nature, because sin is a free and knowing transgression of a divine command. How then can it be that original sin is a free act of the will for us? . . .

St. Thomas (De Malo, q. 4, a. i) says that man must not be treated as a single person but as a member of the human race which has its starting point in Adam, as though all men were a single man. . . . How can they psychologically represent one will in such a way that original sin would become a free act of every member of the race?

The only way of giving a certain answer to this question is to refer back to the pure spirituality of our first parents, a spirituality which would in part have been inherited by their descendants; to

---

[13] Catechism of the Catholic Church, 1994, United States Catholic Conference, Inc.—Libreria Editrice Vaticana § 1039.

the latter there would also have passed that capacity for being influenced, that noopneustia, of which the writer spoke when he showed how angels partake of the knowledge of angels higher than themselves by illumination, and having partaken of that knowledge, obey them. They are influenced with a degree of power which we simply cannot imagine. . . . This noopneustic power rested in Adam who would have been spiritually one with his son (who in his turn would have been similarly one with his own children) and would so have influenced that son that he would have been wholly obedient to his father's will. This will would have been passed on from generation to generation, and would have determined the wills of posterity precisely as the wills of the higher angels determine those of the lower ones—or as the will of the hypnotist influences the will of his subject. Thus we would have been born with the same disposition of will as Adam possessed. . . . Any deviation from this, though physically possible, would have been impossible morally, or would at the most have only been possible in matters of little importance, in so far as this was necessary for the assertion of free will. This accord would have been firm, instantaneous and irrevocable, of the kind we have already noted in the case of pure spirits. Thus the will of posterity was actually contained within the will of Adam, so that his sin became our own, Adam's posterity was infected, "being prone to evil from . . . youth" (Gen. 8. 21) and "sold under sin" (Rom. 7. 14). Adam's sinful act thus became actually morally and psychologically our own. . . .

The dividing wall of individualism was necessarily a consequence of sin. In this way the Catholic doctrine of original sin provides an indication that our first parents, in addition to their human nature, also possessed as the basis of their preternatural gifts that of pure spirits together with all the faculties appertaining to the latter which we have enumerated above. [14]

As mentioned, the original sin affected everyone who will ever live. We also see certain historical figures who have affected history and hence the lives of all who come after them. Examples would be Julius Caesar, George Washington or Adolf Hitler. In one way or another their actions

---

[14] Ibid., Wiesinger, pp. 85-89.

changed the course of societies and hence all who came after them for good or ill.

Just because the things we do in our seemingly little lives are not known as historical events does not mean we do not affect others in profound ways. And, these effects have a ripple effect. The concept of six degrees of separation comes to mind. That says that if you take all the people you know, and all the people each of them knows, and each of them knows and do that six times you will include all the people in the world. This has been tested in a reduced way with the Internet and found to be more or less true.

The point is we are all sort of one, like the fibers in a vast cloth. The fibers at either side of us could be compared to the width of the fabric and are those alive today. Those closest to us we touch. In turn they touch those next to them and so on. But, we are also connected to those behind and before us. That bolt of cloth is the human race. And that single thing is what controls the flow of time. Years ago we had an old oak tree in our back yard that was quite rotten inside. My boys joked that during a strong wind it was only the carpenter ants inside holding hands that kept it from falling down. The human race may be something like that.

The Demands of Nature

If man has the ability, that is the strength, to control the flow of time, and we do it collectively so it is a smooth continuum, is there any reason we do control time rather than letting God do it as part of keeping the physical world in existence?

That man needs time is rather self apparent. There are what philosophers call "the demands of nature" and time would fit into this category.

What is it that we understand by nature and the supernatural? We call all that "natural" which constitutes a substance, or derives from it or which demands it. This means:

1. All that inwardly constitutes the specific essence of a thing, whether it be an essential or an integrating part of its being.

2. Everything that proceeds spontaneously from the nature of a thing, such as aptitudes, talents and powers, and everything that can proceed from it under the influence of some other being, such as proficiency in some art, skill or craft.

3. Everything which, while lying outside the thing itself, is nevertheless necessary for its continued existence (nourishment, light,

air), for its activity (the God-given will for survival), for its development (instruction, society, state) and for the attainment of its goal (knowledge of God, free will). The theologians group all these together under the term "demands of nature" or of **things due, the things that God had to allow men to have, assuming that he desired to create men at all**. Emphasis added. [15]

One could certainly include time in category 3 above. In this case, time would be "necessary for its [the human race's] continued existence," but it says nothing about the control of it any more than we control light or air.

The question is, can time be included in either category 1 or 2? If it can, than the control of time would seem to be one of the "demands of nature" for man. Let's look at category 1 first. It seems time constitutes an "essential or integrating part of its being" for man. But, this says nothing about the control of time. Looking to category 2, is the control of time one of those things that "proceeds spontaneously from the nature of a thing [that is, us] such as "aptitudes, talents and powers. . . ?" If it is, it would come under the category of powers rather than aptitudes or talents to say nothing of an art, skill or craft. If we have the power to control time, this definition is silent about it.

The whole point here is God created us in the odd way that he did. We need time, so he created time as well as the rest of the physical world. So as not to slip into an existential frame of mind, there is a real world out there and there is real time; that is, it exists independently of us.

However, there are small parts of that physical world out there which we can control as when we build a house with a fireplace to make it warm inside when it is cold outside. We do not make the energy that is produced in the fireplace by the burning wood and contained around us by the walls and roof of the house. Yet, even though we do not make the energy, we do exert control over the flow of some of it to keep us warm. The problem with time is we cannot control the flow of a small part of it. It seems to be all or nothing. As far as we know, time is a single thing for everyone. It is not a matter that some people have a lot of it and others very little at a particular instant, i.e. the present. This is not to be confused with the "length of time allotted to each individual" analogous to a piece of string: some people have a much longer piece than others. The point being that the string is the same thickness for all. And, the flow of

---

[15] Ibid., Wiesinger, p 75.

time is the same for all.

If there were only one human he could be expected to cause the flow of time to be rather erratic, as we sense such a great variance of the flow of time as in we have no sense of it at all when in a deep sleep. Other than that, it is not likely that the single person, if there were only one left, could measure the erratic flow of time. But, with the collective sense of the flow of time of billions of people the average is consistent.

From the above it seems we can say that being able to control the flow of time is not a "demand of nature" unless . . . .

Controlling the flow of time by people would not be a demand of nature unless nature could not be counted on to do a proper job of it. After all, we have established that time exists *for* us. Now look at the rest of nature that was created for us. But, there are storms, drought, pestilence, earthquakes and all manner of untoward happenings in the physical world that are not beneficial to man. In spite of this, we are still allowed a measure of nutrition, air, light, etc. or there would be no people. But, the way God has set it up, each of us gets these physical needs in different measure—some get sick, some do not, some are well fed, others starve. Using our free will we make an effort to smooth out these differences by planting crops, building houses and doing all the other things an organized society does. But time, as far as it applies to those living at any one instant, appears to affect all the same. So, is time one part of the physical world where God has made an exception—does He let us control it collectively? It could be it is in the control of time that we find the one place where the entire human race is still in harmony after the sin of Adam. It makes a sort of sense that we need something left to us that we all hold in common.

In this line of reasoning we have examples from society. The United States was made up from the beginning of people from around the world. We were called the melting pot. That is, many peoples came here but they all eventually became Americans. Now, we are becoming an extremely diverse society where the various minorities are not melting into American society. Liberals will say that our strength is in our diversity. That is a fundamentally wrong idea. No society can exist unless all members hold at least some things in common. In the past one of our common things was language—English. Another was the Christian religion. Those are now being lost as a common denominator and we are losing our identity, our cohesiveness and our freedom.

On the other hand, one could object that we all do, in fact, get a different measure of time in that we all have different life spans as mentioned above. That could in turn be answered by saying that the other physical needs are not measured by the total amount, say food, one gets in a life time, short or long, but in how much is allocated *as needed*. So, from that point of view we should each get different amounts of time *as needed*. It appears that nature does not do a proper job of controlling the physical world, and we, using our free wills, can control parts of it, so why could not we, using our free wills, control at least parts of time?

Chapter 4

Man Controls the Flow of Time

There is relatively little mentioned about what controls the flow of time. We have mentioned that it seems to flow at a rate that is consistent with the rate of the sensory perceptions of human beings. It is also obvious that subjectively it flows at different rates as in periods of happiness it seems to fly along while during periods of loneliness, sadness or pain it seems to drag. But, for all people taken collectively, it appears to have a steady rate of flow. We will leave out of this discussion the relativistic effects on time postulated by Einstein.

The control of the flow of time is considered here because if it can be shown that people control the flow of time it would go a long way to showing that time only exists for us.

To prove the conjecture that people control the flow of time requires the establishment of three premises. The first is that time only has meaning for human beings and as a result there was no time before the first human, commonly known as Adam, was created, and that there will be no time after the last human dies. That is, time was created, or if one prefers exists, for us. This has already been covered to some extent. The second premise to show is if we want to prove that man controls the flow of time is that he has the power to do so. The third, then, is to show that if time were created for us, and if we have the power to control its flow, that we indeed do collectively control that flow. This proposition is opposed to what everyone has always assumed which is that the control of the flow of time is something God does along with making the sky blue on a clear day. Or to put it another way, is the *control* of time—not just the existence of time—a "demand of nature." That subject will be discussed along with others as we proceed.

We begin by asking the question of whether there is some force or power that regulates the flow of time. Gardeil refers in passing to the fact that there must be such a thing.

> This much, however, may be said. In the matter of time, as in the theory of place, the Aristotelian position presents, no doubt, certain aspects that will not stand up in the light of contemporary scientific thought. But other aspects have proved more durable. The idea that cosmic motion is a unitary system, *or that a regulating principle of time is necessary*—these, if not others, are far from gone. Emphasis added. [1]

Here we will consider the proposition that it is man, the collective human race that is living at any one time, that controls the flow of time. But first it would be good to explore a few simple philosophical principles that would be needed to foster such a claim.

Causes and Agents

In the preceding chapters some concepts were presented with the primarily intent of getting the reader predisposed to think about time. In the following pages those ideas along with others will be presented in the hope of reaching some useful conclusions.

It has been shown above that time is a demand of nature for man. The question is open as to whether man controls the flow of time. Here we will look at time as a cause or an agent to determine whether or not this will shed light on the control of time or even on what time is.

From various places we get the definition of agent as an entity that is capable of action; a person or thing that performs an action or brings about a certain result, or is able to do so; an active force or substance producing an effect as a chemical agent.

As for causes, here are four main types of causes recognized by philosophy: material, formal, efficient, and final. Take a marble statue as an example. The block of marble is the material cause, that is the material out of which the statue will be made. The finished statue is the formal cause as it gives it its form. The sculptor is the agent that produces the

---

[1] Ibid. Gardeil, Vol. II, Cosmology, p. 126.

statue and is called the efficient cause because he produces the form. In the case of the sculptor the final cause is the reason the sculptor made the statue as for example, fame, money, or simply to make something beautiful. In the more general case final cause explains the direction of things to an end. This is called teleology. In only one of the causes do we meet an agent and that is as an efficient cause. The sculptor operates outside the material and the form. To say it another way an *agent* is the capacity to *do* and a *cause* is the capacity to *be.* Yet in normal parlance we think of a cause only as an agent as in, "He caused the auto accident."

Agent Causes

Now, we must investigate agent causes. To begin, we must distinguish between movers and agents. A mover changes a property or accident of an object as in changing its position, velocity or temperature. That is, physical changes are correlated with movers and changes in substance are correlated to agents. As a result when above we identified efficient cause we were in other words saying an agent cause. Frequently movers and agents are used interchangeably but as seen here there is, if we use precise terms, a difference.

> Thus the form or internal substantial structure comes to be in the potentiality of the fundamental material and depends upon the disposition of properties brought about in the antecedent substances upon which the principal agent acts." [2]

In the case of the statue the block of marble was the antecedent substance and the principal agent was the sculptor who brought about a new substance by producing new properties in the marble. That is quite different from changing the position or temperature of the block of marble which would result from a mover.

On a more fundamental level we must understand the reason why movers and agents exist at all.

> Now in some ways the answer seems too obvious to be require being stated, for when we consider the modification of properties we readily understand that *properties do not and cannot bring them-*

---

[2] Richard J. Connell, *Nature's Causes*, New York, Peter Lang Publishing, 1995, p. 140.

selves into existence, take themselves out of existence or vary themselves. A local motion, for instance, does not accelerate itself, a rising temperature does not increase itself, a magnetic field surrounding an iron bar does not bring itself into existence, etc. . . . We may add, too, that materials do not make themselves into plants and animals, which is to say we need a cause to account not only for individuals but ultimately for the origin of biological species. It seems, then, that a question asking why there are movers and agents is not superfluous, for without them the world would be wholly inexplicable and even more absurd that certain existentialists claim. Emphasis in the original.[3]

Self-Evident Propositions

There are several principles of reason that we as humans use in rational thinking. These are called self-evident propositions or sometimes analytic propositions. They cannot be proven, only explained. They are of the nature that if you understand the terms used in the proposition you understand the proposition. For example: The whole is greater than any of its parts. If you understand the words, you understand the truth of the proposition. A list of some self-evident propositions is as follows.

- Principle of contradiction: Anything either exists or does not exist (a thing cannot both be and not be in the same way at the same time).
This one has the following corollaries.
  • Principle of the excluded middle: Anything either *is* or it *is not.*
  • Principle of identity: Whatever is, *is*, and that which is not, *is not.*
  • Principle of difference: *That which is* is not *that which is not.*
- Principle of efficient causality: That which is not from itself is produced by another.
- Principle of finality: Every agent acts on account of an end.
- Principle of reason for being: Whatever is has an explanation or sufficient reason of its being.
- Every agent acts for the sake of an end.
This one also has corollaries.
  • Those things which are unto an end are not good save in relation to that end. For example, a carburetor on a car engine is only good

---

[3] Ibid., pp. 130-1.

when it is used for its intended purpose. It would not make a good stapler nor a good flashlight.
- The imperfect is for the sake of the perfect.
- The lower is for the sake of the higher.
- The part is for the sake of the whole. That is, the carburetor is made for the sake of the engine.
- The whole is greater than any of its parts.
- The greater does not come from the lesser.
- Two things equal to a third are equal to each other.
- No cause can produce an effect beyond its—the cause's—actual capacity.
- Anything received in any way whatsoever is received according to the limitations of that which is received.
- Good is to be sought and evil avoided.
- Two solid objects cannot occupy the same space at the same time.

This last one mentioned is not always included in the list but is a good place to start. When you are out Christmas shopping and there are no parking spaces around the shopping center that one does concern you. In frustration you drive around saying, "I can't find a parking space." That is not philosophically correct. You can find many parking spaces. What you cannot find is one not already occupied by another solid object. And since your car is also a solid object you know that trying to park in one of the spaces already occupied by another solid object will not lessen your frustration. Anyone will immediately accept this self-evident proposition as true.

The reader may ask why he has been subjected to self-evident propositions, something he was not even slightly in the mood for. The reason is this: rather than simply pop one of them out of nowhere this writer thought it would be good to give a short general description of the subject so as not to seem to be inventing an idea to suit his needs. That being the case the self-evident proposition that concerns us here is the Principle of finality: Every agent acts for the sake of an end.

Man Controls the Flow of Time: The Instrumental Agent Case

We could say that time is an agent that operates outside of man, and acts for the sake of an end which is man. It is not an efficient cause—that is agent—as in the case of the sculptor; it is an instrumental agent as in the chisel that the sculptor uses. Now, it would be wrong to say, continuing with the example used earlier, that the engine determines exactly how the carburetor must be designed, though what the carburetor must *do* for the engine will set limits on how it *can* be designed. In similar fashion, what time must do for man sets limits on how time operates just as, from our other example, the sculptor sets limits on how his chisels are designed, and how they are operated, that is, how they are used to form a statue. The chisel is an instrumental agent only insofar as it is moved by a principal agent, the sculptor. Since time is only for man where it is an instrumental agent for him, he becomes the principal agent (efficient cause) moving it and hence controls the rate at which it flows.

The paragraph immediately above is a bit compact so some expansion of the idea of time as an instrumental agent may help make the concept a bit more natural. We quote from Connell the following:

> An artist who sculpts a statue out of wood obviously must employ chisels and mallets and perhaps other instruments without which the statue could not be produced. Each of these has its own design and function, and to the extent that the artist uses them for removing chips or slices of wood, all he does is apply the tool to its own function. But in addition to its application, each tool receives from the hands of the sculptor a modification of its motion through which it is directed to the production of the effect that is more than a separation of a piece of wood from the main block. Clearly the production of a likeness of Socrates in wood or stone presupposes something more than the chisels and mallets with their inherent design; and so we assign the image-effect to the sculptor because only he is proportioned to the effect that comes about. Over and beyond their design and application, the instruments receive special, directed movement through which they produce the likeness of Socrates, the ultimate effect. . . .
>
> The artist first conceives the likeness of Socrates in his mind and imagination and then executes it. The artist is proportioned to the effect and is able to bring about both the effect and the means though intellectual conceptions that direct the productive actions, whereas the instruments by themselves are not so proportioned. And so to repeat: over and beyond the application of their functions, the instrument must receive from the sculptor special motions by which the likeness is produced. Moreover, the special mo-

tions are transient: they are motions that do not permanently modify the design or character of the instrument-recipients, existing in
the instruments only while the artist is acting upon them. Instruments, then, are agents only while they are moved by a principal
agent, and *only when they bring about an effect that is beyond their
own character or nature, their own operational abilities.* Emphasis
in the original. [4]

In the case of time, when it is used as an instrumental agent by humans it
brings about an effect that is beyond its own character or nature which is
awareness in man. This is so because with no time there would be nothing for man to be aware of since the universe would happen in a timeless
instant.

It could be argued that in the above example after the artist finishes
making the statue and lays down his tools the tools still exist. Since time
exists only for man and will go away when he does, it makes time a special kind of instrumental agent. As such, "the special motion (rate of flow
of time in this case) is transient: its motion does not permanently modify
the design or character of the instrument-recipients (time), but exist in
the instrument (time) only while the artist (man) is acting upon it." In this
special case the instrument goes away when the artist (the human race)
stops acting on it.

In all of this there are three things at work: the artist, the tool (instrumental agent) and the work (statue) to be produced. In our case we have
the artist (man), the instrumental agent (time) and the end work which is
each particular life of living humans which all have awareness as an essential quality. To repeat our initial assertion based upon a self evident
proposition: time is an agent that acts for the sake of an end which is that
strange aware thing that exists in space and time called man.

For an analogous example of this where the instrumental agent goes
away when the "artist" stops using it, consider the following. A plumber
needs to solder copper pipes together to form a water tight joint. He uses
a propane torch, or more exactly a propane flame. The heat from the
flame causes the solder to flow and seal the pipe joint. In this case the
instrumental agent, the flame, does not exist until the plumber lights the
propane gas, and it goes away after the valve on the torch is closed. The
flame is also an agent that does no good if the torch with lit flame is left
setting on the workbench. What would be the point! It is only an instru-

---

[4] Ibid., pp. 156-7.

mental agent when the plumber directs the flame to the pipe joint. So, it is reasonable to say that time is an instrumental agent that not only receives its special, and transient, motions from the principle source (man) and that that instrumental agent ceases to exist when the principal agent ceases to use it, i.e. when the human race in this physical world is ended.

Man Controls the Flow of Time:
The Awareness Case

The issue of whether people control the flow of time must be answered because having it flow much slower or faster then it does would not work for us. In what follows the idea of awareness is key so some words should be said about it. To begin with, we do not mean the common use of the word as in someone asking, "Are you aware of the fact that the government is not honest when it publishes the official amount of inflation each year?" Anyone who is not aware of the fact, that is know, that the government lies is probably not human so will not be reading these words.

What we mean here by awareness is a fundamental property of humanity. It is the ability we have to say, "I exist," or "I am this person, and not just any person, but this special person." This awareness is identified closely with if not identical with ego. Animals do not share this capacity with us. By way of illustration there is a large elm tree in the neighbor's yard. It is home to a pair of gray squirrels. Frequently one sees them chasing around the tree as if playing a game of tag. However, that is no game. They are defending their nesting territory. As a somewhat comical example of this one day an albino squirrel that was totally white was in that tree. There was a mad scramble all through the branches as the two chased the one. The stark difference on color made it easy to keep track of the antagonists. The normally colored animals seemed to be saying, "Hey dude, this ain't no integrated neighborhood. Get you tail out of here." But, color had no bearing on what the pair of squirrels were doing.

The reason for the example is to illustrate that neither awareness nor ego has anything to do with the activities of squirrels. It was simply an instinctive reaction to intruders which has survival advantages. A heavy concentration of squirrels would unduly draw predators and just as importantly the pair of squirrels did not want the local food supply to be depleted by interlopers so they would have enough nourishment to bring their offspring to maturity.

However, one would expect many people to say that awareness has nothing to do with the age of the universe and that time was here before people arrived. Their contention would be that on the large scale there are two main controlling time periods which are the day and the year and both are determined by the scale of astronomical bodies. On the subatomic level the various particle motions like the speed of electrons around the nucleus of an atom is another fundamental motion and it too seems to be controlled by the scale of things. Therefore, as life appeared it did so in tune with these natural rhythms. Many plants and animals have a reproductive cycle keyed to the year. Most people hold the position that people, as animals, have their life processes set to the speed of time that was already existing. It is not surprising that they would make that argument.

It is worthy of note that the day and year as experienced on earth are unique to the earth. At the very least it would be an argument that there is not likely to be life elsewhere in the universe because our day and year combination would be exceedingly rare. There are books written that describe dozens of unique things about the earth and missing any one of them would make life impossible. This is called the anthropic principle. Therefore, let's say the earth is unique. That in itself does not mean that there could not have been non-rational life after the unique motions of the earth were established. We must go to something more fundamental.

It comes down to the idea that before the first human there was no reason for anything else to exist. God does not do frivolous things. There would be no reason in His creating anything that did not have free will, that could say yes to Him or no to Him. What purpose would be served by having a universe full of matter and energy that had no choice but to do what God built into them to do. If one knew all the stresses in the earth and the physical laws that controlled them every earthquake could be predicted to the last second. If one knew the precise instincts that controlled the behavior of an animal as well as all of the physical and biological laws at work, the actions of any animal would be known from moment to moment. This results in an entirely determinant universe from start to end unless man with his free will is interjected into it. Why would God make a thing for the sake of having a thing?

God could watch the universe go from the big bang to the final fizzle as all energy and matter ran down to a final lump of entropy. If it took thirteen billion years—this is the accepted time used by many physicists—to get to where we are now, it will take, say, another hundred billion years to reach that final state. But, for God there is no time so it all happens in a timeless instant. What's the point! So, we say again, time is pointless without a free will being.

Therefore time flows at a rate that makes sense to the way God created the first man who was the first material creature that was aware of time because he was the first material creature that was aware at all. *It is man's awareness that makes time anything but a timeless instant.* That means that there was no time and hence no universe before the first rational animal with an "aware" mind existed whenever that happened. If it happened at the time of the big bang then the thirteen billion years really did happen, though, it would seem we should have developed a lot more than we have in all that time. In any case that is nonsense because as mentioned before, the demands of nature for man could not be met in outer space nor in the center of a star. That means that time certainly did not start before a planet identical to ours existed with an environment in all respects similar to the one in which we live and that place contained at least one aware creature.

Time flows for man and not for rocks, trees or animals. So, does the collective human race control the flow of time? Yes, because it is for them and why not have it flow at a comfortable rate for them. It is similar to someone being the only occupant in a room and this room having a thermostat specific to it. Why wouldn't that person set the temperature of the room to that which best suited him? In the case of where there were fifty people in the room one could ask each one what temperature he preferred. With that information a median temperature could be established and used as the most suitable. Notice median would have to be used so half of the people would want the temperature warmer and half cooler than they would prefer. Using an average would be inadvisable because one individual might have suicidal tendencies and say he wanted the temperature at ten-thousand degrees thus making the average temperature 270 degrees.

In the case of time once again one could at first assume there is only one person in the universe. If this were a very energetic person who was always bemoaning the fact that he could not get enough done in a day, and if he could control time, he might slow it down so as to accomplish more. Or if the person were lazy and lamented the long boring day, he might speed up time so his boredom would pass more quickly. But, since there are many people, each pulling in his own way a rate for the flow of time is reached that is usable by everyone.

If this argument is not fully satisfying to the reader, it must be borne in mind that the awareness of man *is* the present. That plays into what was said above and will also be key to what follows.

Man Controls the Flow of Time: The Act and Potency Case

We have now considered two ways that man could be considered as having control over the flow of time—the instrumental agent case, and the awareness case. These have their merits but may leave the reader still uncertain. Here we will consider the proposition from a more fundamental philosophical point of view.

We find the concept of act and potency is one of the most basic, if not the most basic, ideas of metaphysics. They are considered a single idea because it is difficult to think of one without the other. Potency belongs to those primary analogous notions that cannot be defined in the proper sense of the word. As such, one must proceed inductively, by examples of it, as well as by comparing it to its opposite—act. Here will be given treatments of these concepts by contemporary authors.

Actual Being—Potential Being.—Here we have determinants of real being. A real being that exists is *actual* being. A real being that can exist but does not, is *potential* being. In so far as anything exits, it is actual; hence, actuality is a perfection. Insofar as anything existible does not exist, it is potential; hence, potentiality is an imperfection; it is unfulfillment. This is way Aristotle defines God, the Infinite Being, as *Pure Actuality*. The transit from potentiality to actuality is called *becoming* or *motion* or *change*. There are four chief types of change; change of substance or *substantial change* (as from living body to dead body; as from lifeless food to living flesh); change of *quantity* (as growth or diminution); change in *quality* (as from hot to cold, from ignorant to learned); change of *place* or local change or local movement. In point of change we see illustrated the axiomatic truth that nothing *becomes*, nothing passes from potential to actual, except under the influence of what is already actual. Emphasis Glenn's. [5]

Notice that the statement in the paragraph above, "The transit from potentiality to actuality is called *becoming* or *motion* or *change*" is nearly the same as the definition of time: "Now, movement or motion is a matter of "now this—then that"; it is a matter of "before and after." And motion or change, under the aspect of before-and-after, is the basis of *real time.*" [6]

Another discussion of potency and act is as follows:

The equivalent thought emerges from another Scholastic axiom, that what is in potency cannot be raised to act unless by a being in act. Potency, this means, cannot be raised to the level of act by itself; only a being already in act, exercising efficient causality [efficient cause, one of the four causes discussed above] can bring it about. Yet, though necessarily in act, the agent must also have the potency to act. Is not this a contradiction? No, it is quite possible to be in act and potency at the same time, but not in the same respect. In a created agent act and potency are complementary. For the exercise of efficient causality the agent must be in act through possession of the form (or perfection) which is to be produced in another; and must at the same time be in (active) potency as regards the action to be performed. So, for instance, the intellect when actuated (or in act) by the impressed species, is in (active) potency as regards the act of intellection. [7]

That is to say, when the intellect is actuated by the impressed species of the image of a tree, it is in potency as regards the act of forming the notion of treeness.

Both excerpts contain the fundamental axiom: *potency cannot be raised to the level of act by itself, but only a being already in act, exercising efficient causality, can bring it about.* Act must come before potency or no potency would ever manage by itself to proceed into act. It also means that the being in act must be proportioned to the effect, that

---

[5] Ibid., Glenn p. 234-5.
[6] Ibid., p. 264.
[7] Ibid., Gardeil, Vol. IV, Metaphysics, p.194-5.

is, the act, it wishes to bring from potency. These statements will have a direct bearing on the case to be made for time as we shall see.

To better understand act and potency several short quotations will be presented.

The co-principles of matter and form are distinguished as act (form) and potency (matter). [8]

The statue did not exist in act in the naked marble, but it could be hewn because it was there in potency. In the fabrication it went from a statue in potency to a statue in act. . . . Change, every change, we shall find, is a going from being in potency to being in act. [9]

. . . now, whatever the instance, we find one thing common to the state of potency, namely relation to act. Potency always expresses relation to act. [10]

Act is prior to potency, hence not necessarily attended by a potency. [11]

Thus, in the composition of act and potency act is related to potency as the limited to the limiting. . . . Consequently perfection, is limited by a principle distinct from yet united with it; this principle is potency. Thus, in every composition of act and potency, act is limited by potency. [12]

Potency denotes capacity for perfection, whereas act of its very nature signifies perfection; such opposite notions must correspond to entities that are really distinct. [13]

Act and potency are not two beings but principles of being which determine each other but which while really distinct, do not have each a distinct existence. [14]

---

[8] Ibid., p. 183.
[9] Ibid., p. 186
[10] Ibid., p. 188.
[11] Ibid., p. 191.
[12] Ibid., p. 196
[13] Ibid., p. 197.
[14] Ibid., p. 198.

Joseph P. Cody

It is possible to find philosophers that differ on nearly any point one wants to make. However, those that start with the Greeks and continue to the Scholastics in working out the truth of reality, one finds commonality.

> Substance is a potency in respect to accidents it can receive. An essence is a potency to the act of existence it receives. . . . All these are mixed potencies; that is, they are actual in some respect, but in potency in another respect. Primary matter, on the other hand is pure potency. It is actuated by a substantial form, and with it constitutes a substance. [15]

Notice that in the quote above "An essence is a potency to the act of existence it receives. . . ." Time has the parts of future, present and past. Here we are interested primarily in the future and present. In the above quote "An essence is a potency to the act of existence it receives. . . ." Or the future is in potency to the act of the present it receives.

> . . . an act can receive a further act to which it is in potency, as a substance which receives accidents—for example, water is actual, but in potency of heat that it can receive. But to assert that the same thing can be potency and act in the same respect is to assert the contradictory. It would mean saying that to be heatable (not hot) and to be hot are identical. . . .
>
> It follows that two intrinsic principles are needed to account for any finite thing: a principle of act or perfection to explain its actuality, and a principle of potency to explain its limitations and becoming. No act is found to be limited save by potency. To say something is high in the scale of being is to say that it has greater actuality than lesser entities. [16]

From this we see that cold water is in potency of being hot. Hot water is not; it already has that perfection. Perfection can be lost as well as gained—cold water to hot to cold.

The future is in potency of becoming the perfection called the present and then the perfection is lost to the past—cold to hot to cold. But, the

---

[15] John Young, *The Scope of Philosophy*, Leominster, Herefordshire, UK, Gracewing Publishing, 2010, p. 223.
[16] Ibid., pp. 224-5.

future which is in potency to the present cannot become the present except by the influence of something that is already in act (actual), namely the present— *potency cannot be raised to the level of act by itself, but only by something already in act.* And, the awareness of man is precisely that present—that something—that is in act. That means that since man's awareness is the *act* that brings the potency of the future into being as the present, man necessarily makes the present happen and in so doing regulates how fast this happens by his collective awareness.

Chapter 5

The Main Elements of Our Existence

Summary of Where We Are

One could say that Aristotle, St. Augustine, and St. Thomas had some of the best minds the human race has produced. All three agree that there is no past and no future because the past in no more and the future is not yet. They are ambiguous about the present other than to say is has no length. In addition Aristotle and St. Thomas say that time is made up of two elements, one is motion of real things in the real world that exists apart from us, and the other is the numbering of motion that can only be done by intelligence emanating from a spirit namely our souls. St. Thomas goes a step further and says time is a quasi substance by saying it is composed of matter—past and future—and form—the soul's activity of numbering. Modern philosophers add little to this. As we have seen earlier, Gardeil offers a quote from St. Thomas, ". . . time, it seems, can hardly exist without a mind to piece the parts together. . . . that is why the Philosopher says that without a soul there would be no time." [1] In addition he says "The idea that cosmic motion is a unitary system, or that a regulating principle of time is necessary—these if not others, are far from gone." [2]

There will be no attempt to argue with the above conclusions. It does seem a bit odd, though, as far as this author could find that none of these experts seems to have taken the subject any further. That last quote from Gardeil implies that it is the cosmos that regulates time. Why would one assume that the regulation of time is done by matter rather than a bunch

---

[1] Ibid., Gardeil, Vol. II, Cosmology, p. 124.
[2] Ibid., p. 126.

of intelligent, powerful spirits that are connected to that matter, that is, the human race?

Now, it is time to take a further step to see how the physical universe, human spirits, human bodies and time are connected.

Time vs. Energy

When we say time started with man and will end with him, we must include the physical universe as well, because at the end of time there will be "a new heaven and a new earth." As the ancients have said, time is the measure of motion and without a flow of energy there is no motion. So, not only do people control the flow of time, they keep the universe in existence as a secondary cause after divine providence. More on that later.

We could postulate that time might be considered as the flow of energy or at least closely related to energy's flow. The future is the potential energy in the universe, the present is that cusp where some of the potential energy is converted to kinetic energy producing motion, and the past is the spent energy which is called entropy. The instantaneous consumption of potential energy in the form of kinetic energy is what makes all things move hence the premise that the flow of time and the flow of energy are connected. This is not such a profound statement seeing that all thoughtful people consider time the numbering of motion, and motion, or more precisely the change of motion, is the consumption of energy. In fact it is the heart of the subject we call physics.

Potential energy is in potency to becoming kinetic energy and then to entropy. Kinetic energy cannot simply go back to being potential energy and neither, strictly speaking, can cold water go back to being the identically cold water after being heated because energy has been lost to entropy. In this light, kinetic energy could be called *actual energy* or *real energy* as in *real time* as opposed to potential energy (the future) at one end or entropy (the past) at the other.

A distinction must be made in this regard. A body in uniform motion that is not changing either speed or direction does not consume energy. But such a condition is in reality only a theoretical construct because even a rock in interstellar space will be acted upon by the gravity of the nearest galaxy and thereby be constantly changing speed and direction.

So much said. But, it would be wrong to say time was identically equal to energy. Energy is part of the physical universe and as we have seen energy and matter are convertible one to the other. Time is strictly speaking associated with the human spirit. This would mean that time enters into the *control* of the flow of energy while not being part of it.

In addition we must be careful not to fall into the existential mindset of "I think, therefore I am" which says that the physical world out there is nothing but a projection of the mind. The physical world is really out there aside from me. However, as stated before, without rational, aware human beings, there would not be much point of any of it.

It has also been mentioned that most people would say that the natural timing mechanisms of nature—movement of astronomical bodies, vibrations of subatomic particles—were there before us and we picked up those rhythms. But, as we have seen a good case can be made for man as the controlling element for the flow of time. That means those natural rhythms have been dictated by the speed of flow of time we like.

The Structure of Matter

Matter will be discussed in what follows so it is well to see what modern physics can tell us about the subject. Taking the simplest atom, that of hydrogen, we see that its nucleus is made up of a single particle, a proton, and it has one electron "orbiting" around it. Putting the size of these particles and that atom on a scale we can comprehend, we see that if the proton is the size of a B-B and it is placed on home plate of a baseball field, the electron is smaller than a spec of dust in far center field. All the rest is empty space. But it is not entirely empty. It is filled with the force fields caused by the two particles. The force is transmitted by "force carrier particles" such as photons, virtual particles, and gluons. Photons we know about because they are light waves. The other two are something like mathematical constructs nuclear physicists use to make sense of subatomic reactions. As such, they cannot be observed directly because any such attempt changes the processes in which they would be expected to appear.

Nuclear physicists no longer think of an atom like hydrogen as a proton with an electron orbiting around it like a planet about a star. Now, due to quantum mechanics that atom is thought of as a proton with the location of the electron represented by a "cloud" surrounding the nucleus. The cloud is the force field created by the motion of the electron which travels at about two million meters per second which is about a hundredth the speed of light.

Here is it well to think of the hydrogen atom as it really is. Above we compared it to a baseball field to get as sense of its size relative to the size of the proton and the electron but it certainly is not that large. In fact it is very small. It's diameter is 0.1 nano meters or 0.1 x 10^{-9} meters. If it is that small and the electron is traveling at about a hundredth the speed of light it orbits the nucleus about 6 x 10^{17} times a second. [3] At that rate it would appear to be practically everywhere on the sphere around the nucleus all the time thus giving a sense of a solid object but since hydrogen is a gas we would not be aware of it.

However all of the other atoms are similar to hydrogen, though with increasingly more complicated structures as one goes up the periodic table of the elements. This means that all matter contains very little in the way of "stuff" we can actually identify, and a solid wall is not solid at all, but a web of force fields created by the electrons spinning around the nucleus. It is obvious that if the electrons stopped moving the "cloud" of force surrounding the nuclei of atoms caused by the motion of the electrons would disappear and with it matter as we know it.

Continuing this line of thought we see that modern physics has quantized everything, and that is what is called quantum mechanics. Electromagnetic waves have been shown experimentally to have particle properties. As mentioned above these are called photons and they carry differing amounts of energy depending upon their frequencies. All energy is reducible to photons. All of matter can be reduced to molecules, atoms, subatomic particles such as electrons, protons, and neutrons and them still farther into smaller particles. There is a persistent effort by scientists to discover that time is reducible to individual packets somewhat fancifully called the "chronon" which is a hypothetical quantum or particle of time.

Body, Soul, Motion and Time

In what follows soul is to be taken as meaning the spirit that is the soul of a living human being, not the soul after death.

Few modern scientists will go any further than saying that time is the numbering of motion. They shy away from things philosophical or out of hand reject philosophy as being mutually exclusive of science. An example of a modern physicist who is willing to discuss philosophy is V. E. Smith in his book *Philosophical Physics*. In fact, he quotes Aristotle and

[3] These figures are readily available from many sources particularly a text on modern physics, cf.: Kenneth Krane, *Modern Physics*, John Wiley & Sons, Inc., New York, 1983, pp. 133 ff.

St. Thomas frequently, though in the end he says, with most scientists, that science and philosophy are different realms of knowledge.

> The philosophical science of nature does not expect the empiriological physicist to use a philosophical approach in his measurements but only to remember that beyond measurement there is another view of reality which tell us what things are. [4]

Smith seems to be accepting the idea of two-truths though not specifically. He knows that if he rejects two-truths the scientific community will reject him. Two-truths is a stumbling block for modern science since it limits scientists in many ways. It is important to us here because many of the references cited herein are beyond what is commonly called "pure science" and most people would assume that the study of time would be strictly a laboratory endeavor. Therefore a few more words will be devoted to two-truths. In the quote that follows the reader should keep in mind that in the thirteenth century the idea of what we call science, that is rational truth, the organized study of the physical world, was synonymous with and included most of what we call philosophy. Yet today most scientists place virtually all of philosophy in the category of religious truth and "don't you dare bring your religion into my science."

> Thomas [Aquinas] was sent back to Paris in 1269 to challenge Siger of Brabant and his "two truths" doctrine. Siger claimed that rational truth and religion truth were two separate truths, which could contradict each other. The individual was to keep each in a separate compartment of this brain, as it were, calling upon each as needed and not worrying about any contradictions. The "two truths" was in utter opposition to all that Thomas Aquinas wrote and taught and he proceeded to demolish Siger of Brabant's ideas. God is the source of all truth; He is in fact infinite truth. God reveals religious truths to us directly, and He permits us to determine natural truths through natural reason. But, God is the source of both. Therefore, contradictions cannot possibly both be true. [5]

---

[4] Vincent Edward Smith, *Philosophical Physics*, New York, Harper Brothers, 1950, p. 393.
[5] Anne W. Carroll, *Christ the King Load of History*, Rockford, Illinois, Tan Books, 1994, pp. 186-7.

To continue with Smith, he states many of the same ideas as do our references above concerning time. In fact, he has a Chapter of 39 pages titled "Time: The Measure of Motion." He has little that is new in all those pages stating again and again the definition of the ancients namely that "Time is the number of motion, according to a *before* and *after*." Emphasis his. [6]

At one point he may have inadvertently fallen into writing something rather profound for a scientist to say, though as we have seen not at all unusual for philosophers.

But the real *now* as opposed to the *nows* of reason is the factor in the philosophical physics of time that must receive the heaviest accent. In the real and precise sense, the *now* is a moving boundary, like a point originating or terminating a portion of a line, and when taken in this terminal sense as a limit, the *now* is not a part of time because it is indivisible. Emphasis his. [7]

The part of interest is "the *now* is not a part of time because it is indivisible." First of all the now can hardly be excluded from time because it is the only real part of time there is; it is the part in which we live. But more importantly is his conclusion that the *now* is indivisible. In the references cited earlier the general conclusion is that the present has no length. But, to say the *now* is indivisible, which something that has no dimensions would logically be, implies a little different meaning. It suggests a connection to another thing that is indivisible, namely a spirit.

In the short section on spirit above we saw that a spirit is simple which means it has no parts. It is, therefore as we have observed, indivisible. Human beings are the only creatures that live in the *now* of time. Therefore, they live in this indivisible instant. And *only spirits who are indivisible themselves could live in such a place*. But place is the wrong word because place implies extension and hence has parts—front/back, top/bottom. It is reasonable that spirits live in a "place" or rather a "condition" that is indivisible.

One can see how an indivisible now could be associated with and indivisible spirit, but where does that leave the body which is extended in space? We are each a substance composed of a material body and a spiritual soul. As such the soul is the form for the formless matter of the

---

[6] Ibid., p. 365.
[7] Ibid. p. 368.

body. As with all substances it is the form that makes the primal matter into a thing. In our case that which constitutes the form is a spirit making humans metaphysically different from all other substances. And, even though the spirit has no parts it animates the entire body to the last cell. In that way our spirits control matter by keeping our bodies alive. In addition they control the motion of physical matter when we perform conscious actions as mentioned above.

In our earthly lives the present is ever changing. For pure spirits, if time is a part of them, the present is not changing. In the case of God, He always existed in an unchangeable perfect form. The angels may have had a condition immediately after they were created when time for them was changing and in that time they were given the chance to accept God, or reject Him. After that, time froze for them and they were eternally good or bad. Since God made us in His image and likeness we could assume that means we have time built into us in a fashion somewhat similar to Him and the angels. Only for us, the present keeps changing as long as we live, but will eventually be stopped in eternal bliss or eternal pain where time will be the same for us as it is for God and the angels.

It should be mentioned for completeness that there it the intermediate stage of purgatory. All books on purgatory mention it as extended in time, though a time that flows, or seems to flow to those there, at a different rate from that which we experience. The following account is one of several reported in the literature about purgatory.

Two religious of eminent virtue vied with each other in leading a holy life. One of them fell sick and learned in a vision that he would soon die and remain in Purgatory only until the first Mass should be celebrated for the repose of his soul. Full of joy at these tidings, he hastened to import them to his friend, and entreated him not to delay the celebration of the Mass which was to open Heaven to him.

He died the following morning and his holy companion lost no time in celebrating the Holy Sacrifice. After Mass, whilst he was making his thanksgiving, and still continuing to pray for his departed friend, the latter appeared to him radiant with glory, but in a tone sweetly plaintive he asked why the one Mass which he stood in need had been so long delayed. "My blessed brother," replied the Religious, "I delayed so long, you say? I do not understand you." "What! Did you not leave me to suffer for more than a year before offering Mass for the repose of my soul?" "Indeed, my dear brother, I commenced Mass immediately after your death, not a quarter of an hour has elapsed." [8]

We return now to the concepts of potency and act. Recall that only a thing in act can bring a thing in potency into act. And the agent bringing the potency into act must be proportioned to the effect that comes about. For example, a gallon of water at room temperature is in potency of being in act as boiling water. A red hot paper clip is in act of being hotter than boiling water and it contains heat which is the commodity that is required to heat the water. But, its heat is not proportioned to the heat needed to bringing a whole gallon of water to the boiling point.

Looking at the present as time in act and the future as time in potency we see that the present has what is needed to bring the future into the present somewhat analogous to the heat in the paper clip. But is it proportioned to what is needed to bring the potency of the future into the act of the present? Or as in the case where we postulated that time is the flow of energy, is the kinetic energy of the present proportioned to bring the potential energy of the future into kinetic energy. One could say no because there is no real connection between potential energy and kinetic energy as far as the latter causing the former to become kinetic. It is like the water that is still above the dam in relation to the water that is in the process of falling having tipped over the rim of the dam. The water that is falling has lost all connection to the water still above the dam. For the water still above the dam it requires something completely other than the water to cause more water to fall like, for example, the assistance of gravity. The water already falling is not what causes more water to fall.

That would mean that the present, as such, has no connection with the future. It requires something beyond time to cause some part of that supply of future to become the present. That something beyond time is a spirit, and the only spirit that is intimately connected to both matter and time is the human soul.

To recapitulate, in his section on time H. D. Gardeil is either quoting Aristotle and St. Thomas Aquinas directly or is summarizing their thoughts. "Though time is not motion it is nevertheless unseverable from motion. Take away all change and motion and time disappears." [9] In ad-

---

[8] Fr. F. X. Schouppe, S.J., *Purgatory*, Rockford, Illinois, Tan Books and Publishing, Inc., 1973, p. 64. First published 1926.
[9] Ibid., Gardeil Vol. II, Cosmology, p. 121.

dition, "Aristotle allows that time in its full extension cannot exist apart from mind." [10] And again, "And that is why the Philosopher says that without a soul there would be no time." [11]

We can now make a reasonable further statement. Without a soul there is no time so with no supply of time to draw upon there could be no motion. And, with no motion there would not likely be any matter. With all solid matter being little more than force fields made up of virtual particles, as discussed above, if these particles were in some way nullified or their motion stopped the force fields they create would disappear and the universe with them. That is why we can say that the human soul, motion, matter and time are all mutually necessary. Take away the soul and you take away time. Take away time and motion is not possible. Take away motion and matter goes away and with it the universe—*no soul no time; no time no motion; no motion no matter; no matter no universe.* This goes to a previous point that said the universe does seem to be here specifically for us.

The above paragraph states the case negatively. Speaking more positively we can say that we know of only two exemplars of existing things that have no dimensions, the indivisible present and the indivisible spirit. These have what could be called a mode of existence that is different from matter. A dimensionless present cannot control matter that is spread out in space; it is not proportioned to the task. It requires a dimensionless thing that has been created specifically for the task of interacting with, and moving, spread out matter. That special thing is the human soul. Having the capability of associating intimately with matter, the soul necessarily must extend its powers to the other entity that makes the changing world of matter possible. That is, of course, time.

As mentioned before, if a person decides in the morning to go shopping in the afternoon it makes a different set of futures coming toward him than if he stays home and cleans his closet. But, it makes the future different for everyone else, too. Certainly the clerk at the store where he might have shopped will have a different afternoon depending on the choice. And, as we have seen the idea of six degrees of separation means that everyone on earth will be affected. Now, all the inhabitants of earth are making similar decisions all the time. This combination of willful, conscious acts is what makes the present and hence the world in which we live. It would not matter if there were only one human in existence,

---

[10] Ibid., p. 123.
[11] Ibid., p. 124.

the effect would be the same. And, no, a squirrel grabbing one nut rather than another is not a conscious act.

Final Step

There is a final step that must be taken. That is to ask the question that if time is here for man, and man controls its flow, how does he from moment to moment change that future waiting to happen into the present where he has his existence? Certainly the power of a spirit is capable of such an act, but how is it shaped, how is it made into the continuum of the present? There are at least two factors to consider. One is the spirit of each person alive and the other is Divine Providence. We each make our own present out of the future that is available by our conscious acts, whereas the flow of time and the control of the universe is done with the collective pressure of everyone else and the omniscience of God.

Here the reader is encouraged to go back and read the short section on triune matter. If the soul can affect matter in the brain so we can move our muscles at the command of our mind, it is reasonable the collective spirits of all who are alive can move the amalgam of matter in the universe in a way suitable to the human race. It would seem that a continuous flow of time requires many people because it flows along when we sleep or even daydream and make no decisions. Yet, as long as we are alive our individual spirits must keep us *in* the present whether or not we are conscious of it. It may be that is the main physical—for lack of a better word—duty of our souls vis-à-vis our individual selves. If we are kept in step with the present it would mean all of our other physical processes such as breathing and heart beat are kept functioning. We finally come to a point where we can form a definition of time.

We started by postulating that the essence of time was an accident of human beings that are alive. And that "An accident is an essence whose nature it is to exist in a substance as in a subject." [12] In addition St. Thomas leaves intact Aristotle's assertion that "without a soul there would be no time" [13] so our definition that follows is not something new but a way of restating long accepted philosophical concepts though we are taking issue with the idea that time is nothing but "before and after."

The essence of time is an accidental essence (attribute) of man's

---

[12] Ibid., Thomas Aquinas, *Thomas Aquinas On Being and Essence*, p. 66.
[13] Ibid., H.D. Gardeil, O. P. *Introduction to the Philosophy of St. Thomas Aquinas*, Vol. II, pp. 123-4.

spiritual soul that exhibits itself when the soul is united with a living body. The indivisible spirit operates on the indivisible present to permit our earthly existence and with it the universe. In our minds we can think about the future and the past but only the present exists.

The reader my object to part of that definition on the grounds that we have all been told the universe is kept in existence by divine providence. That is a true statement, but divine providence also makes use of secondary causes. "The truth that God is at work in all the actions of his creatures is inseparable from faith in God the creator. God is the first cause who operates in and though secondary causes. . . ." [14]

The orderly movements of the planets around the sun are caused by divine providence but we recognize a secondary cause at work, namely gravity. Certainly God could make that happen without gravity but it would leave things not making sense. Likewise, when one starts his car and presses on the accelerator the car moves because the tires do not slip on the pavement—another case of divine providence and yet another example of a secondary cause, namely friction. As seems to be His way, God has made the universe in such a way that if we look long enough we will find an explanation of how things work. It is not unreasonable to say that man as a secondary cause keeps the universe in existence—no man, no time; no time, no motion; no motion no matter; no matter no universe.

There is yet another issue about time and the universe to be mentioned. We start by looking at a sheet of normal copy paper. In the United States that has a width and height of 8-1/2 by 11 inches and it's about 0.004 inches thick. What happens to the width and height of the paper when the thickness is reduced to zero? The sheet of paper disappears and with it width and length. We have already arrived at the conclusion that the present—where the universe exists—has no length. Is that like reducing the thickness of the paper to zero? It would imply that the universe doesn't exist other than by having an infinite number of infinity short presents pieced together by the human spirit. Infinity is a hard word normally attributed only to God and maybe that's a place there divine providence especially comes into play.

To think of this another way consider that we accept as true that a human being is a substance composed of a material body and a spiritual soul and that's that. No one ever seems to question how two so dissimilar "things" could come together. It is normally left to something like God

---

[14] Catechism of the Catholic Church, 1994, United States Catholic Conference, Inc.—Libreria Editrice Vaticana §308.

knows what He's doing so forget about it. Let's be a little discourteous and ask the question. The spirit has no parts and the present being indivisible has no parts either; and the present is where we and the universe exists. It seems not only possible but entirely expected that the human spirit and the partless time/universe are specifically made for each other and naturally come together although with the spirit being the superior member.

There remains the little problem that the human spirit and only a small part of the universe come together in the human being. That leaves the rest of the universe in the hands of God. We must think back to where we considered that after the general judgement there will be a new heaven and new earth and the present earth will have passed away. With no more people like we are now, either God withdraws divine providence from the present earth or people as a secondary cause leave the present earth unattended and it disappears. This would mean that in some way the union of spirit and body has an extended effect on the universe. And as we have shown before, spirits certainly have the power to do that. It might be well to go back and reread the section in spirit in this regard especially the part about the power of the spirit.

If the reader finds all of this a bit much there are several facts, some of which are listed below, that must be taken into account if our world and our lives are to make sense.

- Human beings are made up of a spiritual soul and a material body.
- Those two things are from two different realms of being.
- Humans live in time along with the rest of the universe.
- Time is thought of as having three parts—past, present and future.
- The future does not exist because it is not yet.
- The past does not exist because it in no more.
- That leaves the present as the only part of time that exists.
- The present has that annoying property of having no length—indivisible.
- The spirit is simple in that it has no parts—also indivisible.

It should be noted that in this work we have tried to define how time, so to speak, *operates* and have based the definition above on that premise. Obviously time is a most general word. A standard desk dictionary devotes the better part of a whole page to it. So when the ancients speak of time as "before and after" that is one common expression used to represent time and there is no intent to devalue their work. It does seem that the three main sources referenced, Aristotle, St. Augustan, and St. Thomas, each had weightier concerns in their times than spending an inordinate amount of effort on a non-controversial topic like time.

There is a persistent objection that should be answered. We have said that the universe did not exist until the first human being arrived, and that there was no 13 billion years that scientists say make up the age of the universe. Would those billions of years have existed if there had been rational human beings alive all those years? The answer is yes. However, even the most avid evolutionists admit that the human beings, that is Homo sapiens, have only been around for 300,000 years, and that is being wildly generous. They do admit that human civilizations only started at most 20,000 years ago. Forming communities for mutual help which would lead to civilizations is one of the first things humans do so the 20,000 number would seem to be the more reasonable one.

Addenda

Fossils

Fossils have nothing to do with time. Here will be presented a somewhat whimsical explanation of why we have them. For most of us our whole lives have been permeated with the billions and billions of years needed for the universe and us "to evolve." To have that suddenly taken away may leave some readers feeling a bit quaky inside. So, if there were no time and no universe before the first man and woman, why are there fossils? In a simple sentence, the fossils were placed in the earth as a temptation to man. In what follows an effort will be made to make sense out of the statement.

In Genesis we read that God created the universe and man in six days. The evolutionist will say that if someone wants to hold that position, fine, but the six days are not six literal days as we know them, but six epochs. Divide up the thirteen billion years since the big bang into six chunks of whatever length suits your fancy. Others insist that the six days were really six days as we now know them. Why not, they say, God is all powerful and can do anything he wants. It's difficult to argue with that.

There is a third way, though, and that starts with God deciding to make man composed of a material body and a spiritual soul. Having decided that, he naturally had to provide a place for him to live, that is, provide the *demands of nature*, a subject already covered. That being the case, in a blink of His eye there was the universe complete with the earth containing the Garden of Eden and the fossils. To make that concept understandable to the people of his time Moses used the metaphor of six days in the first chapter of Genesis.

The first man and woman were placed in the garden oblivious of the fossils and were told they could live in happiness and contentment with no scorchingly hot days or blizzards, no hunger, no pain. They would work but work would be enjoyable. If they became tired before the work was complete they could simply call upon some of the preternatural gifts they possessed and complete the work by a sort of psychokinesis with no physical effort on their part. After a certain period in the garden they could proceed on to heavenly bliss without suffering death. The only proviso was that there was the somewhat innocuous tree in the center of the garden, the fruit of which they were not to eat. That was the only downside to paradise.

Leave it to a woman, but you guessed it, along came a wondering snake and she took up a conversation with it. In due course both the woman and her mate ate of the forbidden fruit and in consequence were thrown out of paradise.

Notice that, though the fruit of the forbidden tree was appealing, Eve was not drawn to eat of it out of hunger. There were all the other trees of which they could eat. She was drawn to eat the forbidden fruit out of what we could call a sense of ego, of wanting to be and know things she had no need to know. However, to vastly compress Christian theology, there was after the fall a way left for them to get to heaven. That was to work by the sweat of their brows while leading virtuous lives, die, atone for sins not yet atoned for and then ascend to heavenly bliss for eternity. Through the ensuing millennia after their expulsion from paradise there was a running battle with hunger and, in fact, extinction. They mourned the loss of paradise, and gave not a thought to the fossils under their feet except for philosophers here and there, but the concept had no effect on society as a whole. It is of note that the ancient Greeks carried on a rather lively discussion about fossils and evolution so it is not a modern idea as some believe.

Eventually the collective knowledge of the race of humans arrived at what could be called a technological state. One could say it started with Copernicus and Galileo in the sixteenth and seventeenth centuries, or with Newton in the eighteenth. But from there on, the running battle with hunger began to lessen. Technology from the breeding of plants and animals to advanced farm implements made life more secure. In fact, man was on his way to recreating the paradise he had lost and be free from want and privation. But God is no dummy—omniscience precludes that—so he placed a time bomb in the earth, namely those fossils. As seems to be His way, God demands that there is always a decision to be placed before man—to accept God or to reject God.

Along with technology in practical areas of making life easier and better, came knowledge in the life sciences. That in itself was not bad, but it inexorably had to lead to people wondering where life came from and, though it took a leap of faith, the fossils were made to appear as a map to the past all the way to where the first life popped into existence from non-life.

However, if we as a race had kept our eye on the ball, technology would have left free time for the average person to reflect on this life and the life hereafter and use his leisure time to improve his chances of success vis-à-vis heaven or hell. Even in view of our fallen nature if we maintained a knowledge of God a sense of shame and guilt would have remained. But the temptation inserted into the created world by those fossils came to the fore. It made it possible to explain away God and, hence, creation by God and replace it with the flow of evolution in which man was becoming ever more perfected with each passing generation, obvious facts not withstanding. The normal fickleness of man by itself would not have been sufficient to cause the mass apostasy as has happened; it took a general concept that could be made to look as natural as breathing. Enter evolution based on fossils.

That is the temptation posed by the fossils. At the point were the human race could be living in a relatively comfortable state under divine guidance, modern man ate of the forbidden fruit in his modern paradise.

The breach between God and man caused by eating the modern fruit of evolution is not in a strict sense the same as the original breach. Man has been redeemed and still maintains free will today as he did, say, a thousand years ago so any individual is capable of pulling himself at least partially out of this present fallen state with God's help. The difficulty lies in a type of mass amnesia that leaves people unable to think. It is no idle musing to wonder if it is possible for the human race as a whole to shuffle off the coil of narcissistic atheism which has been poured into the psyche of modern man by the incessant barrage of the evolutionists.

Is there a Fossil Record?

Above we mentioned fossils but not a fossil record. That is so because fossils are readily found in most places on the earth but the question remains as to whether or not they form any sort of a record. Here, of course, most people think of *fossil record* as a sort of map showing how various species morphed into existence from prior species as the prior species died out.

There is little that can *honestly* be said about the fossil record. What follows is a quote from Jerry Adler writing in *Newsweek Magazine* about a conference in mid-October, 1980, held at Chicago's Field Museum of Natural History. The majority of 160 of the world's top paleontologists, anatomists, evolutionary geneticists and developmental biologists supported some form of this theory of "punctuated equilibria."

> The missing link between man and the apes . . . is merely the most glamorous of a whole hierarchy of phantom creatures. In the fossil record, missing links are the rule. . . . The more scientists have searched for the transitional forms between species, the more they have been frustrated. . . . Seventy years after quantum theory revolutionized physics, an oddly analogous change has occurred in the theory of evolution and it is just beginning to filter down to public understanding. Evidence from fossils now points overwhelmingly away from the classical Darwinism which most Americans learned in high school: that new species evolve out of existing ones by the gradual accumulation of small changes, each of which helps the organism survive and compete in the environment. Increasingly, scientists now believe that species change little for millions of years and then evolve quickly, in a kind of quantum leap—not necessarily in a direction that represents an obvious improvement in fitness. The theory is still being worked out. Among other points of contention, it is uncertain whether the leap takes place in a few generations or over tens of thousands of years. [1]

Also commenting on the conference mentioned above J. W. G. Johnson has this to say:

> Having admitted that the evidence for evolution is missing, these experts are now working on a theory of evolution without evidence. They are suggesting evolution by huge jumps which would leave no fossil evidence. [2]

The book *Evolution?* by Johnson is one of many good works that takes to task all of the tenets of evolution and demonstrates by presently known facts that they are false.

---

[1] Jerry Adler, *Is Man a Subtle Accident?* Newsweek Magazine, November 3, 1980.

[2] J. W. G. Johnson, *Evolution?*, Australia, 1982, p. 13.

Punctuated equilibrium has as its basis the unspoken assumption that there must be something greater than simple nature to produce new species. Most people who believe in evolution also believe in pantheism whether they admit it or not. Pantheism says that nature is god, and conversely that god is nature. That being the case there is no limit to what a cloud of interstellar gas, or a pile of rocks can do—they are a god.

Dinosaurs

There are many who will object to the whole idea that fossils are just something left lying about as God created the universe when He created the first rational beings so it is reasonable to ask if the dinosaurs really did exist. The direct answer is that dinosaurs could have lived but it would have to have been while people were alive. The idea that the universe is thirteen billion years old is based on the big bang theory which says that it all started from a small speck of super dense energy. The explosion sent the primitive universe out in all directions. All of this is supposedly determined by the light from distant objects being red shifted in proportion to their distance from the observer as a result of the Doppler effect. Evolutionists assume it took thirteen billion years because a lot of time is needed for evolution. However, there are other explanations for the red shift and there is much evidence for a younger and smaller universe than is commonly depicted, the elucidation of which, is beyond our present scope.

With that said, it is possible that when the first rational human beings appeared dinosaurs did roam the earth. This might be indicated by soft tissue found in dinosaur bones that are supposedly millions of years old. Another example of a young earth is pieces of fired pottery and gold jewelry that were found in a coal seam also supposedly millions of years old. These artifacts were present when the coal was formed; they were not interjected later. There are many other such examples, but the evolutionists reject them as outliers. They are not outliers, they are facts that must be accommodated to any sane theory.

Bibliography

—A—

Adler, Jerry, *Is Man a Subtle Accident?* Newsweek Magazine, November 3, 1980.

St. Thomas Aquinas, *Thomas Aquinas On Being and Essence*, translated by Armand Maurer, 2nd Edition, 1968, The Pontifical Institute of Mediaeval Studies, Toronto, Canada.

St. Augustine, *The Confessions of St. Augustine*, translated by Albert C. Outler, Ph.D., D.D., Philadelphia, Westminster Press, 1955.

St. Augustine, *The Confessions of St. Augustine*, translated by Rex Warner, New York, Mentor Books, 1963.

—C—

Carroll, Anne w., *Christ the King Load of History*, Rockford, Illinois, Tan Books, 1

Catechism of the Catholic Church, 1994, United States Catholic Conference, Inc.—Libreria Editrice Vaticana

Clark, Kenneth, *Civilization*, Harper and Row Publishers, NY, 1969.

Connell, Richard J., *Nature's Causes*, New York, Peter Lang Publishing, 1995.

Connell, Richard J., *Substance and Modern Science*, Notre Dame, Indiana, University of Notre Dame Press, 1988.

—G—

Gamow, George, *One, Two, Three... Infinity*, New York, Mentor Books, 1947.

Gardeil, H. D., O.P., *Introduction to the Philosophy of St. Thomas Aquinas,* Vol. II, Cosmology, B. Herder Book Co. English translation 1958.

Gardeil, H. D., O.P., *Introduction to the Philosophy of St. Thomas Aquinas*, Vol. IV, Metaphysics, B. Herder Book Co., 1967.

Glenn, Paul J., *An Introduction To Philosophy*, St. Louis, Mo., B. Herder Book Company, 1943.

—J—

Johnson, J. W. G., *Evolution?*, Australia, 1982.

—K—

Krane, Kenneth, *Modern Physics*, John Wiley & Sons, Inc., New York, 1983.

—L—

Lubar, Henni de, *Catholicism, Christ and the Common Destiny of Man*, London, Bruns and Oats, 1950.

—M—

Maurer, Armand, *Thomas Aquinas On Being and Essence* 2nd Edition, The Pontifical Institute of Mediaeval Studies, Toronto, Canada, 1968.

—O—

Outler, Albert C., Ph.D., D.D., *The Confessions of St. Augustine*, Philadelphia, Westminster Press, 1955.

Bibliography

—P—

Philberth, Bernard, *Des Dreieine* Christiana Verlag, Stein a. Rhein, 1974.

Pontynen, Arthur and Rod Miller, *Western Culture At The American Crossroads*. Wilmington, Delaware, ISI Books, 2011.

—S—

Schouppe, Fr. F. X., S.J., *Purgatory*, Rockford, Illinois, Tan Books and Publishing, Inc., 1973, p. 64. First published 1926

Sheed, Frank J., *Knowing God: God and the Human Condition*, New York, Sheed and Ward, Inc., 1966.

Sheed, Frank J., *Theology for Beginners*, New York, Sheed & Ward, 1957.

Siegfried, Ernst, MD, *MAN The Greatest of Miracles*, Collegeville, MN, The Liturgical Press, 1976.

Smith, Vincent Edward, *Philosophical Physics*, New York, Harper Brothers, 1950.

https://www.space.com/what-is-the-planck-time

—W—

Warner, Rex, *The Confessions of St. Augustine*, New York, Mentor Books, 1963.

Wiesinger, Alosi, O.C.S.O., *Occult Phenomena*, Westminster, Maryland, The Newman Press, 1957.

—Y—

Young, John, *The Scope of Philosophy*, Leominster, Herefordshire, UK, Gracewing Publishing, 2010.

Index

A

accelerator 61
accident 3, 5-7, 37, 60, 68, 71
accidental 6, 60
accidents 2, 7, 48
accumulation 68
Adam 17, 19, 25, 27-29, 32, 35
Adam's 29
Adler, Jerry 68, 71
agent 36-42, 45, 46, 58
agents 36-38, 41
amalgam 60
American 32, 73
Americans 32, 68
amnesia 67
analytic propositions 38
anatomists 68
angelic 25
angels 2, 17, 18, 25, 29, 57
antecedent 37
anthropic 43
apes 68
apostasy 67
Aquinas, St. Thomas 2, 3, 5, 55, 58, 60, 71, 72
Aristotelian 36
Aristotle 1, 3-7, 9, 45, 51, 54, 58, 59, 63
Aristotle's 3, 8, 60
astronomical 43, 53
atheism 67
atom 22, 43, 53
atomic 26

atoms 13, 26, 54
Augustine, St. 1, 3, 5-9, 11, 14, 16, 51, 71-73
Australia 68, 72
axiom 46

B

baseball 53
B-B 53
before-and-after 5, 46
being 1, 2, 4, 5, 7, 8, 11, 12, 15, 18, 19, 22-24, 29-32, 37-39, 44-49, 52-54, 58-63, 65, 68, 69, 71, 72
beings 3, 7, 11, 18, 22, 35, 47, 53, 56, 60, 62, 63, 69
Bible 19
big bang 12, 16, 44, 65
billion 16, 44, 63, 65, 69
biological 38, 43
body-free 24
Buddhism 13

C

Caesar 29
Calhoun, Lake 12
carburetor 38-40
Carroll, Anne W. 55, 71
Catechism 28, 61, 71
Catholic 28, 29, 61, 71
Catholicism 13, 72
Christendom 12

75

lied 12
lifeless 45
limit 56, 69
logical 5, 18
Loon Lake 12

M

magnetic 26, 38
marble 36, 37, 47
mathematical 53
matter 2, 5, 12, 14, 16, 18, 21-23, 25-27, 31, 36, 43, 44, 46-48, 51-54, 56-61
measurable 3
measure 5, 32, 33, 52, 56
Mediaeval 2, 71, 72
memorials 12
metaphysically 57
metaphysics 5, 45, 46, 72
microsecond 16
millennia 66
Miller, Rod 73
millions 68, 69
mind 4, 7, 9, 16, 24, 30, 31, 40, 44, 45, 51, 53, 55, 59, 60
Minneapolis 12
modernism 12
molecules 26, 54
moon 5, 17, 23
motion 3-6, 13, 14, 36, 38, 40, 41, 43, 45, 46, 51-54, 56-59, 61
movers 37, 38

N

nano 54
narcissistic 67

nature 1-3, 14, 15, 21, 23, 25, 27-33, 36, 38, 41, 44, 47, 53, 55, 60, 65, 67, 69
neurons 15
neutrons 54
Newton 66
nonbeing 8
noopneustia 29
noopneustic 29
nuclear 18, 26, 53
nuclei 54
nucleus 43, 53, 54
numbering 3, 4, 13, 51, 52, 54
numerable 3, 5, 6

O

occult 23, 24, 28, 73
occults 23
offspring 21, 43
omniscience 60, 66
Omniscient 27
ontologically 23
orbits 54
organism 27, 68
outliers 69

P

paleontologists 68
pantheism 69
paradise 66, 67
paranormal 24
particle 26, 43, 53, 54
particles 26, 53, 54, 59
partless 22, 23, 62
periodic table 54
phantasm 15
phantom creatures 68
phenomena 23, 24, 28, 73

star 44, 53
stars 5, 13, 17
statue 36, 37, 40, 41, 47
subatomic 26, 43, 53, 54
substance 2-4, 13, 30, 36, 37,
    45, 48, 51, 56, 60, 61, 71
supernatural 11, 30
supernovae 18

T

talents 30, 31
technological 66
technology 66, 67
teleology 37
telos 13
temporal 7
theologians 28, 31
theology 13, 22, 66, 73
Thomas, St. 1-9, 28, 51, 55, 58,
    60, 63, 71, 72
thoughts 11, 13, 15, 17, 19, 23,
    26, 58
time 1-9, 11-19, 21, 25, 27, 30-
    33, 35-49, 51-63, 65-67, 69
timeless 41, 44
tool 40, 41
torch 41
transit 5, 45, 46
trinitarian 26
triune matter 26, 27, 60

U

uncountable 27
unitary system 36, 51
universe 1, 12, 13, 16-18, 22,
    27, 41, 43, 44, 52, 59-63, 65,
    69
unscientific 26

V

vegitation 19
velocity 37
virtuous 66

W

war 12, 18
waves 26, 53, 54
Wiesinger, Alois 23-25, 29, 31,
    73
will-power 26
wills 21, 29, 33
world 2, 5, 7, 12, 13, 17, 18, 21,
    23, 25, 27, 28, 30-33, 38, 42,
    51, 53, 55, 59, 62, 67

Y

Young, John 48, 73

Z

zero 16, 61